INFINITE GROUP RINGS

PURE AND APPLIED MATHEMATICS

A Series of Monographs and Textbooks

COORDINATOR OF THE EDITORIAL BOARD

S. Kobayashi

UNIVERSITY OF CALIFORNIA AT BERKELEY

1. Kentaro Yano. Integral Formulas in Riemannian Geometry (1970)
2. S. Kobayashi. Hyperbolic Manifolds and Holomorphic Mappings (1970)
3. V. S. Vladimirov. Equations of Mathematical Physics (A Jeffrey, editor; A. Littlewood, translator) (1970)
4. B. N. Pshenichnyi. Necessary Conditions for an Extremum (L. Neustadt, translation editor; K. Makowski, translator) (1971)
5. L. Narici, E. Beckenstein, and G. Bachman. Functional Analysis and Valuation Theory (1971)
6. D. S. Passman. Infinite Group Rings (1971)

In Preparation:

W. Boothby and G. L. Weiss (eds.). Symmetric Spaces: Short Courses Presented at Washington University

Y. Matsushima. Differentiable Manifolds (E. J. Taft, editor; E. T. Kobayashi, translator)

L. Dornhoff. Group Representation Theory

INFINITE GROUP RINGS

DONALD S. PASSMAN

Department of Mathematics
University of Wisconsin
Madison, Wisconsin 53706

MARCEL DEKKER INC. New York 1971

MARCEL DEKKER, INC.
270 Madison Avenue, New York, New York 10016

LIBRARY OF CONGRESS CATALOG CARD NUMBER: 72-163311

ISBN: 0-8247-1523-3

Current printing (last digit):
10 9 8 7 6 5 4 3 2

PRINTED IN THE UNITED STATES OF AMERICA

TO
Marjorie

PREFACE

The group ring $K[G]$ is an associative ring which exhibits properties of the group G and the field of coefficients K. As the name implies, its study is a meeting place for group theory and ring theory, and as such it has been approached from many different points of view. For example, the finite group theorist does character theory and the analyst does operator theory and Fourier analysis. On the other hand, our approach here is purely algebraic and ring theoretic.

This algebraic study of group rings was initiated in 1949 by I. Kaplansky, but it did not really catch on until the fundamental work of S. A. Amitsur appeared some ten years later. Since then the subject has been pursued by a small but growing number of researchers, and at this point in time it has reached the stage in its development where a coherent account of the basic results is needed. Such is the goal of this text, and a rough outline of the topics covered is given below.

In Chapter I we study linear identities in $K[G]$ and find one basic method yields a number of diverse results. Thus we obtain necessary and sufficient conditions for $K[G]$ to be prime or semiprime and we obtain partial results on the possibility of $K[G]$ satisfying a polynomial identity. The study of this latter problem is continued in Chapter II but is there reformulated in terms of bounded representation degree. The main result here settles the case in which $K[G]$ is semiprime.

In Chapter III we offer a number of conditions which guarantee the semisimplicity of $K[G]$. We also completely describe the structure of its nilpotent radical and we obtain necessary and sufficient conditions for $K[G]$ to have a maximal nilpotent ideal. Chapter IV begins with a study of the trace of idempotents in characteristic zero and the support of central idempotents in general. Next we consider the possibility of the group ring being regular and we study the structure of annihilator ideals.

Finally we list and discuss a number of open problems in the field. While many of these are of the "famous unsolved" variety, others most certainly can be done and some may even be easy.

Needless to say, we have been somewhat selective in the topics covered here. For one thing, we study neither group rings over arbitrary rings of coefficients nor twisted group rings since these generalizations merely add unnecessary complications. For another, we have tried to choose material which offers a nice balance between the ring theory and the group theory used. Therefore we make no claim to completeness.

We have however attempted to make this work as self-contained as possible, and a basic one year course in algebra should provide sufficient background knowledge. In particular, the reader is assumed to be familiar with basic group theory (subgroups, conjugacy classes, Sylow theorems), basic noncommutative ring theory (Wedderburn theorems, Jacobson radical, Density theorem), and field theory (Galois theory, separable and inseparable extensions).

I would like to express my appreciation to a number of people: Richard Brauer, who first interested me in group rings; Martin Isaacs, my good friend and collaborator in a number of papers; S. A. Amitsur and David Wallace, whose work in the field stimulated my own research; Martha Smith, whose recent thesis is a major breakthrough in the study of polynomial identities; and I. N. Herstein, who offered me a week's vacation in Chicago. Much of the research for this book was done at the Institute for Defense Analyses, Princeton, New Jersey where I spent a year's leave of absence from the University of Wisconsin. I would like to thank Richard Leibler, Director of IDA-CRD, for supplying me with a nice quiet place to work, most of the time.

Madison, Wisconsin D. S. Passman
April, 1971

CONTENTS

I

LINEAR IDENTITIES

§1. BASIC REDUCTION

Let K be a field and let G be a (not necessarily finite) group. We let $K[G]$ denote the group ring of G over K; that is, $K[G]$ is a K-algebra with basis $\{x \mid x \in G\}$ and with multiplication defined distributively using the group multiplication in G.

If $\alpha = \sum k_x x \in K[G]$, we define the support of α to be

$$\operatorname{Supp} \alpha = \{x \in G \mid k_x \neq 0\}.$$

Then $\operatorname{Supp} \alpha$ is a finite subset of G.

Suppose for a moment that α is central in $K[G]$ and let $x \in \operatorname{Supp} \alpha$. If $y \in G$, then

$$x^y = y^{-1}xy \in \operatorname{Supp} y^{-1}\alpha y = \operatorname{Supp} \alpha.$$

Since $\operatorname{Supp} \alpha$ is a finite set, it follows that there are only a finite number of distinct x^y with $y \in G$. The set of all elements $x \in G$ with this property will be of great interest to us. We define

$$\Delta = \Delta(G) = \{x \in G \mid [G : \mathbf{C}_G(x)] < \infty\}.$$

Since the conjugates of x are in one-to-one correspondence with the right cosets of $\mathbf{C}_G(x)$, it follows that x has only finitely many conjugates if and only if $x \in \Delta$.

We can now observe that Δ is a normal subgroup of G. First $1 \in \Delta$ and since $\mathbf{C}_G(x^{-1}) = \mathbf{C}_G(x)$ we see that $x \in \Delta$ implies $x^{-1} \in \Delta$. Finally, since a conjugate of xy is the product of a conjugate of x with one of y, it follows that if $x, y \in \Delta$, then $xy \in \Delta$. Thus Δ is a subgroup of G and it is clearly normal. It is called the F. C. (finite conjugate) subgroup of G.

The importance of Δ here is two-fold. First we are frequently able to reduce the problems studied from $K[G]$ to $K[\Delta]$ and second we are able to handle the much simpler group Δ. In this section we consider the reduction to $K[\Delta]$ which will yield results on prime and semiprime group rings.

Lemma 1.1. Let $H_1, H_2, \ldots, H_n$ be subgroups of G of finite index. Then $H = H_1 \cap H_2 \cap \cdots \cap H_n$ has finite index in G and in fact

$$[G:H] \leq [G:H_1][G:H_2] \cdots [G:H_n].$$

Proof. If Hx is a coset of H, then clearly $Hx = H_1x \cap H_2x \cap \cdots \cap H_nx$. Since there are at most $[G:H_1][G:H_2] \cdots [G:H_n]$ choices for $H_1x, H_2x, \ldots, H_nx$, the result follows.

Lemma 1.2. Let G be a group and let $H_1, H_2, \ldots, H_n$ be a finite number of subgroups. Suppose that there exists a finite collection of elements $x_{ij} \in G$ $\big(i = 1, 2, \ldots, n;\ j = 1, 2, \ldots, f(i)\big)$ with

$$G = \bigcup_{ij} H_i x_{ij},$$

a set theoretic union. Then for some i, $[G:H_i] < \infty$.

Proof. By relabeling we can assume all the H_i to be distinct. We prove the result by induction on n, the number of distinct H_i. The case $n = 1$ is clear.

If a full set of cosets of H_n appears among the $H_n x_{nj}$, then $[G:H_n] < \infty$ and we are finished. Otherwise if $H_n x$ is missing, then $H_n x \subseteq \bigcup_{ij} H_i x_{ij}$. But $H_n x \cap H_n x_{nj}$ is empty so $H_n x \subseteq \bigcup_{i \neq n, j} H_i x_{ij}$. Thus

$$H_n x_{nr} \subseteq \bigcup_{\substack{i \neq n \\ j}} H_i x_{ij} x^{-1} x_{nr}$$

and G can be written as a finite union of cosets of $H_1, H_2, \ldots, H_{n-1}$. By induction $[G:H_i] < \infty$ for some $i = 1, 2, \ldots, n-1$ and the result follows.

Let θ denote the projection $\theta: K[G] \to K[\Delta]$ given by

$$\alpha = \sum_{x \in G} k_x x \to \theta(\alpha) = \sum_{x \in \Delta} k_x x.$$

Then θ is clearly a K-linear map but it is certainly not a ring homomorphism in general.

Lemma 1.3. Let $\alpha, \beta, \gamma, \delta \in K[G]$ and suppose that for all $x \in G$ we have $\alpha x \beta = \gamma x \delta$. Then $\theta(\alpha)\beta = \theta(\gamma)\delta$ and $\theta(\alpha)\theta(\beta) = \theta(\gamma)\theta(\delta)$.

Proof: We first show that $\theta(\alpha)\beta = \theta(\gamma)\delta$. Suppose by way of contradiction that $\theta(\alpha)\beta \neq \theta(\gamma)\delta$ and let $v \in \operatorname{Supp}(\theta(\alpha)\beta - \theta(\gamma)\delta)$.

Suppose $\operatorname{Supp}\theta(\alpha) \cup \operatorname{Supp}\theta(\gamma) = \{u_1, u_2, \ldots, u_r\}$ and set $W = \bigcap \mathbf{C}_G(u_i)$. Since $u_i \in \Delta$, it follows from Lemma 1.1 that $[G:W] < \infty$.

Write $\alpha = \theta(\alpha) + \alpha'$, $\gamma = \theta(\gamma) + \gamma'$ where $\operatorname{Supp}\alpha' \cap \Delta = \varnothing$, $\operatorname{Supp}\gamma' \cap \Delta = \varnothing$, and then write the finite sums

$$\begin{aligned} \alpha' &= \sum a_i y_i, & \gamma' &= \sum c_i y_i, & y_i &\notin \Delta \\ \beta &= \sum b_i z_i, & \delta &= \sum d_i z_i \end{aligned}$$

with $a_i, b_i, c_i, d_i \in K$ and $y_i, z_i \in G$. If y_i is conjugate to some vz_j^{-1} in G choose $h_{ij} \in G$ with $h_{ij}^{-1} y_i h_{ij} = vz_j^{-1}$.

Let $x \in W$. Then

$$\begin{aligned} 0 &= x^{-1}\alpha x\beta - x^{-1}\gamma x\delta \\ &= \big(x^{-1}\theta(\alpha)x\beta - x^{-1}\theta(\gamma)x\delta\big) + x^{-1}\alpha' x\beta - x^{-1}\gamma' x\delta \\ &= \big(\theta(\alpha)\beta - \theta(\gamma)\delta\big) + x^{-1}\alpha' x\beta - x^{-1}\gamma' x\delta \end{aligned}$$

since $x \in W$ implies that x centralizes $\theta(\alpha)$ and $\theta(\gamma)$. Now v occurs in $\operatorname{Supp}(\theta(\alpha)\beta - \theta(\gamma)\delta)$ and so this element must be canceled by something from the other terms. Thus there exists y_i, z_j with $v = x^{-1}y_i x z_j$ or

$$x^{-1} y_i x = vz_j^{-1} = h_{ij}^{-1} y_i h_{ij}.$$

Thus $xh_{ij}^{-1} \in \mathbf{C}_G(y_i)$ and $x \in \mathbf{C}_G(y_i)h_{ij}$. We have therefore shown that

$$W \subseteq \bigcup_{ij} \mathbf{C}_G(y_i)h_{ij}.$$

Now $[G:W] < \infty$ so if $G = \bigcup Ww_k$, then by the above we have

$$G = \bigcup_{ijk} \mathbf{C}_G(y_i)h_{ij}w_k,$$

a finite union of cosets. By Lemma 1.2, $[G:\mathbf{C}_G(y_i)] < \infty$ for some i, a contradiction since $y_i \notin \Delta$. Thus $\theta(\alpha)\beta = \theta(\gamma)\delta$.

Now write $\beta = \theta(\beta) + \beta'$, $\delta = \theta(\delta) + \delta'$ where $\operatorname{Supp}\beta' \cap \Delta = \varnothing$, $\operatorname{Supp}\delta' \cap \Delta = \varnothing$. Then

$$\begin{aligned} 0 &= \theta(\alpha)\beta - \theta(\gamma)\delta \\ &= \big(\theta(\alpha)\theta(\beta) - \theta(\gamma)\theta(\delta)\big) + \big(\theta(\alpha)\beta' - \theta(\gamma)\delta'\big). \end{aligned}$$

Since $\operatorname{Supp}(\theta(\alpha)\theta(\beta) - \theta(\gamma)\theta(\delta)) \subseteq \Delta$ and $\operatorname{Supp}(\theta(\alpha)\beta' - \theta(\gamma)\delta') \cap \Delta = \varnothing$, we clearly have $\theta(\alpha)\theta(\beta) - \theta(\gamma)\theta(\delta) = 0$ and the result follows.

Theorem 1.4 (Passman [*35*]). Let A and B be ideals in $K[G]$. Then

(i) $\theta(A)$ is an ideal in $K[\Delta]$.
(ii) $A \neq 0$ if and only if $\theta(A) \neq 0$.
(iii) $AB = 0$ implies that $\theta(A)\theta(B) = 0$.

Proof. (i) Since $\theta(\alpha_1) + \theta(\alpha_2) = \theta(\alpha_1 + \alpha_2)$, $\theta(A)$ is clearly closed under addition. Furthermore, if $\alpha \in A$ and $\gamma \in K[\Delta]$, then $\alpha\gamma \in A$, $\gamma\alpha \in A$, and we have easily $\theta(\alpha\gamma) = \theta(\alpha)\gamma$, $\theta(\gamma\alpha) = \gamma\theta(\alpha)$. Thus $\theta(A)$ is an ideal in $K[\Delta]$.

(ii) Certainly $\theta(A) \neq 0$ implies that $A \neq 0$. Now suppose $A \neq 0$ and let $\alpha \in A$, $\alpha \neq 0$. If $x \in \operatorname{Supp}\alpha$, then since A is an ideal, $x^{-1}\alpha \in A$ and $1 \in \operatorname{Supp} x^{-1}\alpha$. Thus $0 \neq \theta(x^{-1}\alpha) \in \theta(A)$ and $\theta(A) \neq 0$.

(iii) Finally let $\alpha \in A$, $\beta \in B$. If $x \in G$, then $\alpha x \in A$ so $\alpha x\beta \in AB = 0$ and $\alpha x \beta = 0$. By Lemma 1.3 with $\gamma = \delta = 0$, we have $\theta(\alpha)\theta(\beta) = 0$ and thus $\theta(A)\theta(B) = 0$.

§2. PRIME RINGS

A ring R is said to be prime if for any two ideals A, B in R, $AB = 0$ implies $A = 0$ or $B = 0$. In this section, we consider the possibility of $K[G]$ being prime. We start by studying $\Delta(G)$.

Lemma 2.1. Let G be a group with a central subgroup Z of finite index. Then G', the commutator subgroup of G, is finite.

Proof. Let $(x, y) = x^{-1}y^{-1}xy$ denote commutators in G. Since $(x, y)^{-1} = (y, x)$, we see that G' is the set of all finite products of commutators and it is unnecessary to consider inverses.

Let $x_1, x_2, \ldots, x_n$ be coset representatives for Z in G and set $c_{ij} = (x_i, x_j)$. We observe first that these are all the commutators of G. Let $x, y \in G$ and say $x \in Zx_i$, $y \in Zx_j$. Then $x = ux_i$, $y = vx_j$ with u and v central in G. This yields easily $(x, y) = (x_i, x_j) = c_{ij}$.

Now let $x, y \in G$. Since Z is normal in G and G/Z has order n, we have $(x, y)^n \in Z$. Thus

$$\begin{aligned}(x, y)^{n+1} &= x^{-1}y^{-1}xy(x, y)^n = x^{-1}y^{-1}x(x, y)^n y \\ &= x^{-1}y^{-1}x(x^{-1}y^{-1}xy)(x, y)^{n-1}y \\ &= x^{-1}y^{-2}xy^2 \cdot y^{-1}(x, y)^{n-1}y \\ &= (x, y^2)(y^{-1}xy, y)^{n-1}\end{aligned}$$

since conjugation by y being an automorphism of G implies that

$$y^{-1}(x, y)^{n-1}y = (y^{-1}xy, y^{-1}yy)^{n-1} = (y^{-1}xy, y)^{n-1}.$$

We show finally that every element of G' can be written as a product of at most n^3 commutators and this will yield the result. Suppose $u \in G'$ and $u = c_1 c_2 \cdots c_m$, a product of m commutators. If $m > n^3$, then since there are at most n^2 distinct c_{ij}, it follows that some c_{ij}, say $c = (x, y)$, occurs at least $n + 1$ times. We shift $n + 1$ of these successively to the left using

$$\begin{aligned}(x_r, x_s)(x, y) &= (x, y)c^{-1}(x_r, x_s)c \\ &= (x, y)(c^{-1}x_r c, c^{-1}x_s c)\end{aligned}$$

and obtain $u = (x, y)^{n+1}c'_{n+2}\, c'_{n+3} \cdots c'_m$ where each c'_i is a possibly new commutator. Using

$$(x, y)^{n+1} = (x, y^2)(y^{-1}xy, y)^{n-1}$$

we can then write u as a product of $m - 1$ commutators. Thus every element of G' is a product of at most n^3 of the c_{ij} and thus clearly G' is finite.

Lemma 2.2. Let H be a finitely generated subgroup of $\Delta(G)$. Then $[H:\mathbf{Z}(H)]$ and $|H'|$ are finite. Thus if $\Delta(G)$ contains no nonidentity elements of finite order, then $\Delta(G)$ is torsion free abelian.

Proof. Let H be generated by $x_1, x_2, \ldots, x_n$. Since each x_i has only a finite number of conjugates in G, they have only a finite number of conjugates in H. Hence $[H:\mathbf{C}_H(x_i)] < \infty$. By Lemma 1.1, $Z = \bigcap \mathbf{C}_H(x_i)$ has finite index in H. Since $x_1, x_2, \ldots, x_n$ generate H, we see that Z is central in H. Thus $[H:\mathbf{Z}(H)]$ is finite and by Lemma 2.1, H' is finite.

Now suppose $\Delta(G)$ has no nontrivial elements of finite order and let $x, y \in \Delta(G)$. Set $H = \langle x, y\rangle$. Since H is a finitely generated subgroup of $\Delta(G)$, the above implies that H' is finite and hence $H' = \langle 1\rangle$. Thus x and y commute and $\Delta(G)$ is abelian. By definition $\Delta(G)$ is torsion free.

Lemma 2.3. Group G has a finite normal subgroup H whose order is divisible by a prime p if and only if $\Delta(G)$ contains an element of order p.

Proof. Let H be given. Since p divides $|H|$, H contains an element x of order p. Since H is normal in G, all conjugates of x are contained in H and hence $x \in \Delta$.

Now let $x \in \Delta$ have order p. Let $x_1 = x, x_2, \ldots, x_n$ be the finite number of distinct conjugates of x. If $H = \langle x_1, x_2, \ldots, x_n \rangle$, then $H \subseteq \Delta$ and H is normal in G since conjugation by an element of G merely permutes the generators of H. By Lemma 2.2, H' is finite. Now H/H' is a finitely generated abelian group generated by elements of finite order. Thus H/H' is finite and H is finite. Since $x \in H$, p divides the order of H and the result follows.

Lemma 2.4. Let H be a torsion free abelian subgroup of G and let $\alpha \in K[H] \subseteq K[G]$ with $\alpha \neq 0$. Then α is not a zero divisor in $K[G]$.

Proof. We show that $\alpha\beta = 0$ implies that $\beta = 0$. An analogous proof works in the other direction. Suppose $\alpha\beta = 0$. We can choose $y_1, y_2, \ldots, y_k$ in distinct right cosets of H in G so that $\beta = \beta_1 y_1 + \beta_2 y_2 + \cdots + \beta_k y_k$ with $\beta_i \in K[H]$. Then

$$0 = \alpha\beta = (\alpha\beta_1)y_1 + (\alpha\beta_2)y_2 + \cdots + (\alpha\beta_k)y_k$$

and since $\alpha\beta_i \in K[H]$ we have clearly $\alpha\beta_i = 0$. Thus it suffices to show that $\alpha\beta_i = 0$ implies $\beta_i = 0$ or equivalently we can assume that $G = H$ is a torsion free abelian group.

Assume then that $G = H$. Now there certainly exists a finitely generated subgroup $W \subseteq G$ with $\alpha, \beta \in K[W]$. Thus we may also assume that $G = W$ is finitely generated. By the fundamental theorem of abelian groups $G = \langle x_1 \rangle \times \langle x_2 \rangle \times \cdots \times \langle x_n \rangle$, a finite direct product of infinite cyclic groups. Then $K[G]$ is essentially a polynomial ring in the variables $x_1, x_2, \ldots, x_n$ except that negative exponents are also allowed. It is now obvious that $K[G]$ is an integral domain so $\alpha\beta = 0$ implies $\beta = 0$.

Theorem 2.5 (Connell [*11*]). The following are equivalent.

(i) $K[G]$ is prime.

(ii) $\Delta(G)$ is torsion free abelian.

(iii) G has no nonidentity finite normal subgroup.

Proof. (i) $\Rightarrow$ (iii). Suppose G has a nonidentity finite normal subgroup H. Set

$$\alpha = \sum_{x \in H} x \in K[G].$$

Since H is normal in G, $y^{-1}Hy = H$ for all $y \in G$ and thus $y^{-1}\alpha y = \alpha$. Hence α is central in $K[G]$ and clearly $\alpha \neq 0$.

If $y \in H$, then $yH = H$ so $y\alpha = \alpha$. This yields

$$\alpha^2 = \left(\sum_{x \in H} x\right)\alpha = |H|\,\alpha$$

and hence $(\alpha - |H| \cdot 1)\alpha = 0$. Since $H \neq \langle 1 \rangle$, we have clearly $\alpha - |H| \cdot 1 \neq 0$. Set

$$A = (\alpha - |H| \cdot 1)K[G], \qquad B = \alpha K[G].$$

Since α is central, these are both nonzero ideals. Moreover clearly $AB = 0$ so $K[G]$ is not prime, a contradiction. Hence H does not exist.

(iii) ⇒ (ii). By Lemma 2.3, $\Delta(G)$ has no nonidentity elements of finite order and then by Lemma 2.2, $\Delta(G)$ is torsion free abelian.

(ii) ⇒ (i). Let A and B be ideals in $K[G]$ with $AB = 0$. By Theorem 1.4, we have $\theta(A)\theta(B) = 0$ and hence by Lemma 2.4 either $\theta(A) = 0$ or $\theta(B) = 0$. The result follows from Theorem 1.4.

An interesting consequence of the above is

Theorem 2.6 (Connell [*11*]). The group ring $K[G]$ is Artinian if and only if G is finite.

Proof. If G is finite, then $K[G]$ is a finite dimensional algebra and hence clearly Artinian.

Assume now that $K[G]$ is Artinian. Then by Hopkin's theorem, $K[G]$ is also Noetherian and its right regular representation has a composition series. Let $\ell(K[G])$ denote the common length of all such series. We show that G is finite by induction on $\ell(K[G])$. If $\ell(K[G]) = 1$, then 0 is the unique ideal of $K[G]$ so certainly $K[G]$ is prime and we will consider this case later. Let $\ell(K[G]) = n$ and assume the result is true for all groups W with $\ell(K[W]) < n$.

Suppose first that $K[G]$ is not prime. Then by Theorem 2.5, G has a finite normal subgroup $H \neq \langle 1 \rangle$. Since the kernel of the natural epimorphism $K[G] \to K[G/H]$ is a nonzero ideal, it is clear that $K[G/H]$ is Artinian and $\ell(K[G/H]) < \ell(K[G]) = n$. Thus by induction $|G/H| < \infty$. Since $|H| < \infty$, we see that G is finite.

Now suppose that $K[G]$ is prime. Since $K[G]$ is Artinian, this implies that $K[G]$ is semisimple and then by Wedderburn's theorem, that $K[G]$ is simple. Thus the natural epimorphism $K[G] \to K[G/G]$ must be an isomorphism and this implies that $G = \langle 1 \rangle$. The result follows.

It is shown in Ref. [*15*] that if G has a finite subnormal series in which each quotient is either infinite cyclic or finite, then $K[G]$ is Noetherian. However it is not known whether these conditions on G are also necessary for the group ring to be Noetherian.

§3. SEMIPRIME RINGS

Let R be a ring. An ideal P of R is said to be prime if R/P is a prime ring. Thus P is prime if and only if for all ideals $A, B \subseteq R$, we have $AB \subseteq P$ implies $A \subseteq P$ or $B \subseteq P$. R is said to be semiprime if the intersection of all prime ideals of R is 0. In particular, R is semiprime if and only if it is a subdirect product of prime rings.

Lemma 3.1. Ring R is semiprime if and only if R contains no nonzero ideal with square zero.

Proof. Suppose R contains a nonzero ideal A of square 0. If P is any prime ideal in R, then $A \cdot A = 0 \subseteq P$ so $A \subseteq P$. Hence A is contained in the intersection of all such prime ideals and R is not semiprime.

Now suppose that R contains no nonzero ideal of square 0. Let $\alpha \in R$, $\alpha \neq 0$. We define a sequence $T = \{\alpha_1, \alpha_2, \ldots, \alpha_n, \ldots\}$ of nonzero elements of R inductively as follows. First $\alpha_1 = \alpha$. Second, given $\alpha_n \neq 0$, then the ideal $R\alpha_n R$ does not have square 0. Thus for some $\beta_n \in R$, we have $\alpha_n\beta_n\alpha_n \neq 0$. Set $\alpha_{n+1} = \alpha_n\beta_n\alpha_n$. Since $0 \notin T$, it follows that T is disjoint from some ideal of R, namely 0. By Zorn's lemma, there exists an ideal P of R maximal with respect to $P \cap T = \varnothing$. We show that P is prime. Let A and B be ideals of R with $A \nsubseteq P$, $B \nsubseteq P$. Then $P + A$ and $P + B$ properly contain P, so by the maximality of P it follows that for some i, j we have $\alpha_i \in P + A$, $\alpha_j \in P + B$. If $m = \max(i, j)$, then clearly $\alpha_m \in P + A$, $\alpha_m \in P + B$ so

$$\alpha_{m+1} = \alpha_m\beta_m\alpha_m \in (P + A)(P + B) \subseteq P + AB.$$

Since $\alpha_{m+1} \notin P$, we have $AB \nsubseteq P$ and P is prime. Since $\alpha = \alpha_1 \notin P$, the result follows.

An element $\alpha \in R$ is said to be nilpotent if $\alpha^n = 0$ for some positive integer n. An ideal I of R is nil if all elements of I are nilpotent.

Theorem 3.2 (P. Jordan). Suppose that K is a subfield of the complex numbers which is closed under complex conjugation. Then $K[G]$ contains no nonzero nil ideal.

Proof. Let * denote complex conjugation and extend * to a map of $K[G]$ to itself by

$$\alpha = \sum_{x \in G} k_x x \to \alpha^* = \sum_{x \in G} k_x^* x^{-1}.$$

Clearly $(\alpha^*)^* = \alpha$ and $(\alpha\beta)^* = \beta^*\alpha^*$. In addition, the coefficient of $1 \in G$ in $\alpha\alpha^*$ is $\sum_{x \in G} |k_x|^2$ and thus $\alpha\alpha^* = 0$ if and only if $\alpha = 0$.

Let I be a nil ideal in $K[G]$ and let $\alpha \in I$. Since I is an ideal, we have $\alpha\alpha^* \in I$ and hence for some $n \geq 1$, $(\alpha\alpha^*)^n = 0$. Let n be minimal with this property. Suppose that $n > 1$ and set $\beta = (\alpha\alpha^*)^{n-1}$. Clearly $\beta = \beta^*$ so we have $\beta\beta^* = (\alpha\alpha^*)^{2n-2} = 0$ since $2n - 2 \geq n$. Thus $\beta = 0$ by the above, contradicting the minimality of n. This shows that $n = 1$, $\alpha\alpha^* = 0$, and hence $\alpha = 0$. Thus $I = 0$.

We will show in Chapter III that $K[G]$ has no nonzero nil ideal for any field K of characteristic 0.

Theorem 3.3 Let K be a field of characteristic 0. Then $K[G]$ is semiprime.

Proof. Suppose $K[G]$ is not semiprime. Then by Lemma 3.1, $K[G]$ contains a nonzero ideal A with $A^2 = 0$. Let $\alpha = \sum_{i=1}^{n} k_i x_i \in A$, $\alpha \neq 0$, and let F be the subfield of K generated over the rationals by $k_1, k_2, \ldots, k_n$. Then $F[G] \subseteq K[G]$ and $A \cap F[G]$ is a nonzero ideal in $F[G]$ of square zero. Thus it clearly suffices to assume that $K = F$ or equivalently that K is finitely generated over the rationals. This implies that K is contained in the complex numbers C and we fix an embedding. Then $K[G] \subseteq C[G]$ and AC is a nonzero ideal of $C[G]$ with square zero. This is a contradiction by Theorem 3.2 and the result follows.

We now consider fields of characteristic $p > 0$. Let R be a ring. We set $[R, R]$ equal to the set of all finite sums of Lie products $[\alpha, \beta] = \alpha\beta - \beta\alpha$ with $\alpha, \beta \in R$.

Lemma 3.4. Let E be an algebra over a field K of characteristic $p > 0$ and let k and n be positive integers. If $\alpha_1, \alpha_2, \ldots, \alpha_n \in E$, then

$$(\alpha_1 + \alpha_2 + \cdots + \alpha_n)^{p^k} = \alpha_1^{p^k} + \alpha_2^{p^k} + \cdots + \alpha_n^{p^k} + \beta$$

for some $\beta \in [E, E]$.

Proof. Observe that

$$(\alpha_1 + \alpha_2 + \cdots + \alpha_n)^{p^k} = \alpha_1^{p^k} + \alpha_2^{p^k} + \cdots + \alpha_n^{p^k} + \beta$$

where β is the sum of all words $\alpha_{i_1}\alpha_{i_2}\cdots\alpha_{i_{pk}}$ with at least two distinct subscripts occurring. If words ω_1 and ω_2 are cyclic permutations of each other, that is, if

$$\omega_1 = \alpha_{i_1}\alpha_{i_2}\cdots\alpha_{i_{pk}}$$
$$\omega_2 = \alpha_{i_j}\alpha_{i_{j+1}}\cdots\alpha_{i_{pk}}\alpha_{i_1}\cdots\alpha_{i_{j-1}}$$

then $\omega_1 - \omega_2 = \gamma\delta - \delta\gamma \in [E, E]$ where

$$\gamma = \alpha_{i_1}\alpha_{i_2}\cdots\alpha_{i_{j-1}} \quad \text{and} \quad \delta = \alpha_{i_j}\alpha_{i_{j+1}}\cdots\alpha_{i_{pk}}.$$

Hence modulo $[E, E]$, all cyclic permutations of a word are equal. For convenience we let the cyclic group Z_{p^k} act on the set of these words by performing the cyclic shifts. Then the number of formally distinct permutations of a word ω occurring in β is the size of a nontrivial orbit of Z_{p^k} and hence is divisible by p. Since K has characteristic p, the result follows.

Lemma 3.5. Let K be a field of characteristic $p > 0$ and let $\alpha \in K[G]$ be nilpotent. If $1 \in \operatorname{Supp} \alpha$, then there exists $x \in \operatorname{Supp} \alpha$ such that $x \neq 1$ and the order of x is a power of p.

Proof. If $\beta = \sum b_x x \in K[G]$, then we set $\tau(\beta) = b_1$, the coefficient of 1. τ is clearly a K-linear map of $K[G]$ onto K. Now $[K[G], K[G]]$ is spanned over K by all Lie products of the form $[x, y]$ with $x, y \in G$. Furthermore, if $\tau([x, y]) \neq 0$, then certainly $y = x^{-1}$ and then $[x, y] = xx^{-1} - x^{-1}x = 0$, a contradiction. Hence $\tau([K[G], K[G]]) = 0$.

Write $\alpha = k_1 1 + k_2 x_2 + \cdots + k_n x_n$ where $k_i \in K$ and the x_i are distinct nonidentity elements of G. By assumption, $k_1 \neq 0$. Since $\alpha^m = 0$ for some $m > 0$, it follows that $\alpha^{p^k} = 0$ for some integer $k > 0$. By Lemma 3.4

$$0 = \alpha^{p^k} = (k_1 1)^{p^k} + (k_2 x_2)^{p^k} + \cdots + (k_n x_n)^{p^k} + \beta$$

where $\beta \in [K[G], K[G]]$. Since $0 = \tau(0) = \tau(\beta)$ and $\tau((k_1 1)^{p^k}) = k_1^{p^k} \neq 0$, we conclude that for some $i = 2, 3, \ldots, n$, $\tau((k_i x_i)^{p^k}) \neq 0$. Thus $k_i \neq 0$, $x_i \neq 1$, and $x_i^{p^k} = 1$. The result follows.

Theorem 3.6 (Passman [*35*], Connell [*11*]). Let K be a field of characteristic $p > 0$ and let G have no elements of order p. Then $K[G]$ has no nonzero nil ideals.

Proof. Let I be a nontrivial nil ideal in $K[G]$ and let $\beta \in I$, $\beta \neq 0$. If $y \in \operatorname{Supp} \beta$, then $\alpha = y^{-1}\beta \in I$ and $1 \in \operatorname{Supp} \alpha$. Since α is nilpotent, we conclude from Lemma 3.5 that there exists $x \in \operatorname{Supp} \alpha$, $x \neq 1$ such that x has order a power of p. Thus G has an element of order p, a contradiction.

The converse to the above is decidedly false. As we will see later, there are many examples of groups G with elements of order p such that $K[G]$ has no nontrivial nil ideals.

Theorem 3.7 (Passman [*35*]). Let K be a field of characteristic $p > 0$. The following are equivalent.

(i) $K[G]$ is semiprime.

(ii) $\Delta(G)$ has no elements of order p.

(iii) G has no finite normal subgroup with order divisible by p.

Proof. (i) $\Rightarrow$ (iii). Suppose G has a finite normal subgroup H whose order is divisible by p. Set

$$\alpha = \sum_{x \in H} x \in K[G].$$

As in the proof of Theorem 2.5, we see that $\alpha \neq 0$, α is central in $K[G]$, and $\alpha^2 = |H|\,\alpha$. Now $p \mid |H|$ and K has characteristic p, so $|H| = 0$ in K. Thus if $A = \alpha K[G]$, then A is a nonzero ideal of $K[G]$ and $A^2 = 0$. By Lemma 3.1, $K[G]$ is not semiprime, a contradiction. Hence H does not exist.

(iii) $\Rightarrow$ (ii). This follows from Lemma 2.3.

(ii) $\Rightarrow$ (i). Let A be an ideal in $K[G]$ with $A^2 = 0$. Then by Theorem 1.4, $\theta(A)$ is an ideal in $K[\Delta]$ with $\theta(A)^2 = 0$. Now Δ has no elements of order p so by Theorem 3.6, $\theta(A) = 0$. Hence by Theorem 1.4 we have $A = 0$ and $K[G]$ is semiprime by Lemma 3.1.

§4. POLYNOMIAL IDENTITIES

Let $K[\zeta_1, \zeta_2, \ldots]$ be the polynomial ring over field K in the noncommuting indeterminates $\zeta_1, \zeta_2, \ldots$. An algebra E over K is said to satisfy a polynomial identity, if there exists $f(\zeta_1, \zeta_2, \ldots, \zeta_n) \in K[\zeta_1, \zeta_2, \ldots]$, $f \neq 0$, with

$$f(\alpha_1, \alpha_2, \ldots, \alpha_n) = 0$$

for all $\alpha_1, \alpha_2, \ldots, \alpha_n \in E$. For example, any commutative algebra satisfies $f(\zeta_1, \zeta_2) = \zeta_1\zeta_2 - \zeta_2\zeta_1$.

The standard polynomial of degree n is defined by

$$[\zeta_1, \zeta_2, \ldots, \zeta_n] = \sum_{\sigma \in S_n} (-1)^\sigma \zeta_{\sigma(1)} \zeta_{\sigma(2)} \cdots \zeta_{\sigma(n)}.$$

Here S_n is the symmetric group of degree n and $(-1)^\sigma$ is 1 or -1 according as σ is an even or an odd permutation. We will also use $s_n(\zeta_1, \zeta_2, \ldots, \zeta_n)$ to denote this polynomial.

In this section, we will prove the famous theorem of Amitsur and Levitzki on the polynomial identities satisfied by K_m, the ring of $m \times m$ matrices over K.

Lemma 4.1. Suppose E is an algebra over a field K which satisfies a nontrivial polynomial identity of degree n. Then E satisfies the polynomial identity $f \in K[\zeta_1, \zeta_2, \ldots, \zeta_n]$ with

$$f = \sum_{\sigma \in S_n} a_\sigma \zeta_{\sigma(1)} \zeta_{\sigma(2)} \cdots \zeta_{\sigma(n)}$$

where $a_\sigma \in K$ and they are not all zero.

Proof. A monomial in $K[\zeta_1, \zeta_2, \ldots]$ is an element of the form $\zeta_{i_1}\zeta_{i_2} \cdots \zeta_{i_r}$. These of course form a basis for $K[\zeta_1, \zeta_2, \ldots]$ over K.

Let $g = g(\zeta_1, \zeta_2, \ldots)$ be the given polynomial of degree n satisfied by E. Suppose some variable ζ_i occurs in some but not all the monomials in the expression for g. Then $g = g' + g''$ where ζ_i occurs in all the monomials of g' and in none of g''. Then $g'' \neq 0$, degree $g'' \leq n$, and

$$g''(\zeta_1, \zeta_2, \ldots, \zeta_i, \ldots) = g(\zeta_1, \zeta_2, \ldots, 0, \ldots)$$

so g'' is also clearly a polynomial identity for E. Thus by replacing g by g'' we have a polynomial identity for E of degree less than or equal to n which is a function of fewer variables. We continue in this manner until we obtain a nonzero polynomial h of degree less than or equal to n with the property that each variable ζ_i which occurs in h in fact occurs in each monomial. Since degree $h \leq n$, we see that h is a function of at most n variables. By changing notation if necessary, we may assume that $h \in K[\zeta_1, \zeta_2, \ldots, \zeta_n]$.

Let $\mathscr{H}$ be the set of all $h \in K[\zeta_1, \zeta_2, \ldots, \zeta_n]$, $h \neq 0$, which are polynomial identities for E of degree less than or equal to n and for which all variables which are involved in h occur in each monomial. By the above, $\mathscr{H}$ is nonempty. We choose $f \in \mathscr{H}$ to be a function of the maximal possible number, say t, of variables. We show now that f has the desired property.

Suppose that some monomial in f is not linear in say ζ_1. Since degree $f \leq n$ and $f \in \mathscr{H}$, this implies that f cannot be a function of all the ζ_i, so say, ζ_n is missing. Set

$$f' = f(\zeta_1 + \zeta_n, \zeta_2, \ldots) - f(\zeta_1, \zeta_2, \ldots) - f(\zeta_n, \zeta_2, \ldots).$$

It follows easily that $f' \neq 0$ and that $f' \in \mathscr{H}$. Furthermore f' is a function of $t + 1$ variables, a contradiction. Hence all monomials in f are linear in each variable and thus they all have degree $t \leq n$. If $t < n$, then say ζ_n is missing, and setting $f'' = \zeta_n f$ yields a contradiction. Thus $t = n$ and f has the desired form.

Lemma 4.2. Let $E = K_m$ be the ring of $m \times m$ matrices over K. Then E does not satisfy a polynomial identity of degree less than $2m$.

Proof. Suppose by way of contradiction that E satisfies a polynomial identity of degree $n < 2m$. By Lemma 4.1, we may assume that E satisfies

$$f(\zeta_1, \zeta_2, \ldots, \zeta_n) = \zeta_1 \zeta_2 \cdots \zeta_n + \sum_{\substack{\sigma \in S_n \\ \sigma \neq 1}} a_\sigma \zeta_{\sigma(1)} \zeta_{\sigma(2)} \cdots \zeta_{\sigma(n)}.$$

Let $\{e_{ij}\}$ denote the set or matrix units in E, that is, e_{ij} is the matrix whose only nonzero entry is a 1 in the (i, j)th position. Since $n < 2m$, we may set

$$\zeta_1 = e_{11}, \quad \zeta_2 = e_{12}, \quad \zeta_3 = e_{22}, \quad \zeta_4 = e_{23}, \quad \zeta_5 = e_{33}, \ldots .$$

Then $\zeta_1 \zeta_2 \cdots \zeta_n$ at these values is not zero but clearly for all $\sigma \neq 1$, $\zeta_{\sigma(1)} \zeta_{\sigma(2)} \cdots \zeta_{\sigma(n)}$ at these values is zero. Thus E does not satisfy f.

We now proceed to show that K_m does in fact satisfy a polynomial identity of degree $2m$, namely the standard identity. Let $\{e_{ij}\}$ denote as above a set of matrix units in K_m. We remark that if r is an integer with $1 \leq r \leq m$, then $K_m^{(r)}$, the set of all matrices in K_m whose rth row and rth column are zero, is an algebra isomorphic to K_{m-1}.

Lemma 4.3. The standard polynomial $s_n(\zeta_1, \zeta_2, \ldots, \zeta_n)$ has the following properties.

(i) s_n is linear in each variable.

(ii) $s_n(\ldots, \zeta_j, \ldots, \zeta_i, \ldots) = -s_n(\ldots, \zeta_i, \ldots, \zeta_j, \ldots)$. Thus up to a $\pm$ sign, all orderings of the variables are equivalent.

(iii) $s_n(\ldots, \zeta_i, \ldots, \zeta_i, \ldots) = 0$.

(iv) $s_{n+1}(\zeta_1, \zeta_2, \ldots, \zeta_{n+1}) = \Sigma_i \pm \zeta_i s_n(\zeta_1, \ldots, \hat{\zeta}_i, \ldots, \zeta_{n+1})$. Thus if an algebra E satisfies s_n, then it satisfies s_{n+1}.

(v) $s_{2n}(1, \zeta_2, \zeta_3, \ldots, \zeta_{2n}) = 0$.

Proof. (i) This is clear.

(ii) The map of S_n to S_n given by $\sigma \to \sigma\tau$ where τ is the transposition (i, j) sets up a one-to-one correspondence between the terms of $s_n(\ldots, \zeta_j, \ldots, \zeta_i, \ldots)$ and $s_n(\ldots, \zeta_i, \ldots, \zeta_j, \ldots)$. Under this correspondence, the monomials are the same but the signs differ since $(-1)^{\sigma\tau} = -(-1)^\sigma$.

(iii) We see easily that each monomial in $s_n(\ldots, \zeta_i, \ldots, \zeta_i, \ldots)$ occurs twice, once with a plus sign and once with a minus sign.

(iv) This follows since $\pm\zeta_i s_n(\zeta_1, \ldots, \hat{\zeta}_i, \ldots, \zeta_{n+1})$ is clearly equal to the sum of all terms in $s_{n+1}(\zeta_1, \zeta_2, \ldots, \zeta_{n+1})$ which start with ζ_i on the left. Here of course $\hat{\zeta}_i$ indicates that the variable ζ_i is deleted.

(v) With each ordering of $\zeta_2, \zeta_3, \ldots, \zeta_{2n}$, the element 1 can be placed in precisely $2n$ positions. Half of these terms have plus signs and the other half have minus signs. The result follows.

Lemma 4.4

(i) Let s''' be the sum of all those terms in $s_n(\zeta_1, \ldots, \zeta_n)$ in which the product $\zeta_1\zeta_2\zeta_3$ occurs. Then

$$s''' = s_{n-2}(\zeta_1\zeta_2\zeta_3, \zeta_4, \ldots, \zeta_n).$$

(ii) Let s'' be the sum of all those terms in $s_n(\zeta_1, \ldots, \zeta_n)$ in which the product $\zeta_1\zeta_2$ occurs. Then

$$s'' = \pm s_{n-2}(\zeta_3, \ldots, \zeta_n)\zeta_1\zeta_2 + \sum_3^n (\pm) s_{n-2}(\zeta_1\zeta_2\zeta_i, \zeta_3, \ldots, \hat{\zeta}_i, \ldots, \zeta_n).$$

Proof. (i) It is clear there is a one-to-one correspondence between the monomials which occur in s''' and in $s_{n-2}(\zeta_1\zeta_2\zeta_3, \zeta_4, \ldots, \zeta_n)$. We merely have to check that they occur with the same sign. This sign can of course be found by seeing how many transpositions are required to put the factors of the monomial into their natural order. Consider a term $\zeta_{i_1}\zeta_{i_2}\cdots\zeta_{i_r}\zeta_1\zeta_2\zeta_3\zeta_{i_{r+1}}\cdots$. In s''' or equivalently in $s_n(\zeta_1, \zeta_2, \ldots, \zeta_n)$, the product $\zeta_1\zeta_2\zeta_3$ can be shifted to the left most position by successively shifting ζ_1, ζ_2, and then ζ_2 through the r preceding ζ_i's and this requires $3r$ transpositions. In $s_{n-2}(\zeta_1\zeta_2\zeta_3, \zeta_4, \ldots, \zeta_n)$, $\zeta_1\zeta_2\zeta_3$ is considered as a single entity, and it can be shifted to the left most position with r transpositions. Since $(-1)^{3r} = (-1)^r$, the result follows.

(ii) By (i) above and Lemma 4.3(ii), $\pm s_{n-2}(\zeta_1\zeta_2\zeta_i, \zeta_3, \ldots, \hat{\zeta}_i, \ldots, \zeta_n)$ contains all those terms in s'' with $\zeta_1\zeta_2$ followed by ζ_i. Clearly

$$\pm s_{n-2}(\zeta_3, \ldots, \zeta_n)\zeta_1\zeta_2$$

contains all those terms with $\zeta_1\zeta_2$ followed by nothing.

Lemma 4.5. Let $m > 1$ and assume that K_{m-1} satisfies the polynomial identity s_{2m-2}. Let $\{e_{a_ib_i}\}$ be a set of $2m$ matrix units in K_m and suppose that $s_{2m}(e_{a_1b_1}, e_{a_2b_2}, \ldots, e_{a_{2m}b_{2m}}) \neq 0$. For each $u = 1, 2, \ldots, m$, let $\omega(u)$ equal the number of times u occurs as a subscript in $\{e_{a_ib_i}\}$. Then the possible configurations for the values $\omega(u)$ in some order are

$$3,\ 5,\ 4,\ 4,\ 4,\ 4, \ldots .$$

$$3,\ 3,\ 6,\ 4,\ 4,\ 4, \ldots .$$

or

$$4,\ 4,\ 4,\ 4,\ 4,\ 4, \ldots .$$

Proof. We first show that for all u, $\omega(u) \geq 3$. Suppose first that $\omega(u) = 0$ for some u. Then $\{e_{a_i b_i}\} \subseteq K_m^{(u)}$ and $K_m^{(u)} \simeq K_{m-1}$. By assumption, K_{m-1} satisfies s_{2m-2} and hence also s_{2m} by Lemma 4.3(iv). Since $s_{2m}(e_{a_1 b_1}, e_{a_2 b_2}, \ldots, e_{a_{2m} b_{2m}}) \neq 0$, we have a contradiction.

Now suppose $\omega(u) = 1$ for some u and say e_{iu} is one of the matrix units. It is clear that the only nonvanishing monomials in $s_{2m}(e_{a_1 b_1}, e_{a_2 b_2}, \ldots, e_{a_{2m} b_{2m}})$ must have e_{iu} occurring in the right position. Thus $s_{2m}(\cdots) = \pm s_{2m-1}(\cdots)e_{iu}$. Now all the variable values in $s_{2m-1}(\cdots)$ are contained in $K_m^{(u)}$ and hence as above $s_{2m-1}(\cdots) = 0$, a contradiction. We get a similar contradiction assuming e_{ui} is the matrix unit with subscript u.

Now suppose $\omega(u) = 2$. Clearly $s_{2m}(\cdots) = 0$ if the two subscripts u occur in one of the following three ways: e_{uu}, e_{iu} and e_{ju}, e_{ui} and e_{uj}. The only remaining possibility is that e_{ui} and e_{ju} are the matrix units having u as a subscript. Then the nonzero monomials in $s_{2m}(\cdots)$ can only occur in one of two mutually disjoint ways. First we could have e_{uj} in the left most position and e_{iu} in the right most position. The sum of all such terms of this type is clearly $\pm e_{uj} s_{2m-2}(\cdots) e_{iu} = 0$ since K_{m-1} satisfies s_{2m-2}. Second we could have e_{iu} followed directly by e_{uj}. Since $e_{iu}e_{uj} = e_{ij} \in K_m^{(u)}$, we see that the sum of all such terms of this type is $s'' = 0$ by Lemma 4.4(ii) and the assumption on K_{m-1}. Thus $\omega(u) \geq 3$ for all u.

Since $s_{2m}(\cdots) \neq 0$, some monomial product is not zero. This implies easily that either all $\omega(u)$ are even or all but two are even. Now precisely $4m$ subscripts occur in the set $\{e_{a_i b_i}\}$, so $\sum_1^m \omega(u) = 4m$. These two facts and $\omega(u) \geq 3$ clearly yield the result.

Theorem 4.6 (Amitsur–Levitzki [*6*]). K_m, the ring of $m \times m$ matrices over field K, satisfies the standard polynomial identity of degree $2m$.

Proof. We proceed by induction on m. The case $m = 1$ is clear since K_1 is commutative and $s_2(\zeta_1, \zeta_2) = \zeta_1\zeta_2 - \zeta_2\zeta_1$. Let $m > 1$ and suppose that K_{m-1} satisfies s_{2m-2}.

Let e and f be two orthogonal primitive idempotents in K_m. We show first that $s_{2m}(e, f, K_m, K_m, \ldots, K_m) = 0$. We can of course choose a basis of matrix units so that $e = e_{11}, f = e_{22}$. If $s_{2m}(e, f, K_m, K_m, \ldots, K_m) \neq 0$, then by linearity there exists a set of matrix units $\{e_{a_i b_i}\}$

containing e_{11} and e_{22} with $s_{2m}(e_{a_1b_1}, e_{a_2b_2}, \ldots, e_{a_{2m}b_{2m}}) \neq 0$. By Lemma 4.5 there exists at most one u with $\omega(u) > 4$ and thus either $\omega(1) \leq 4$ or $\omega(2) \leq 4$. Say $\omega(1) \leq 4$. We then replace e_{11} in $s_{2m}(\cdots)$ by $e_{11} = 1 - \sum_2^m e_{ii}$ and use linearity. By Lemma 4.3(v), $s_{2m}(1, \ldots) = 0$. In all the other terms, we now have $\omega(1) \leq 4 - 2 = 2$, so these terms must also vanish by Lemma 4.5, a contradiction.

Let e be a primitive idempotent in K_m. We show now that $s_{2m}(e, K_m, K_m, \ldots, K_m) = 0$. Choose a basis of matrix units so that $e = e_{11}$ and consider $s_{2m}(e_{a_1b_1}, e_{a_2b_2}, \ldots, e_{a_{2m}b_{2m}})$ with $e_{11} \in \{e_{a_ib_i}\}$. If all $e_{a_ib_i}$ have a subscript 1, then since there are at most $2m - 1$ such matrix units, it follows that two entries in $s_{2m}(\cdots)$ are equal and hence $s_{2m}(\cdots) = 0$ by Lemma 4.4(iii). If not then some e_{ij} occurs with $i, j \neq 1$. By the preceding paragraph, $s_{2m}(\cdots) = 0$ if $i = j$, so assume that $i \neq j$. Then $e_{ij} = (e_{ii} + e_{ij}) - e_{ii}$ is a difference of two primitive idempotents each orthogonal to e_{11}. By linearity and the results of the preceding paragraph, we have $s_{2m}(\cdots) = 0$.

Finally we show that $s_{2m}(K_m, K_m, \ldots, K_m) = 0$. By linearity, it suffices to show that $s_{2m}(e_{a_1b_1}, e_{a_2b_2}, \ldots, e_{a_{2m}b_{2m}}) = 0$. Let $e_{ab} \in \{e_{a_ib_i}\}$. If $a = b$, then e_{aa} is a primitive idempotent and $s_{2m}(\cdots) = 0$ by the above. If $a \neq b$, then $e_{ab} = (e_{aa} + e_{ab}) - e_{aa}$ is a difference of two primitive idempotents and hence by linearity we have $s_{2m}(\cdots) = 0$. Thus K_m satisfies s_{2m} and the result follows by induction.

§5. POLYNOMIAL IDENTITY RINGS

In this section, we begin our study of group rings which satisfy a polynomial identity. The first result gives a sufficient condition for this to occur.

Theorem 5.1 (Kaplansky [*26*], Amitsur [*3*]). Let group G have an abelian subgroup A with $[G:A] = n < \infty$. Then $K[G]$ satisfies the standard polynomial identity of degree $2n$.

Proof. Let $x_1, x_2, \ldots, x_n$ be a set of right coset representatives for A in G. Let $E = K[A]$ and $V = K[G]$. Then clearly V is a left E-module with basis $\{x_1, x_2, \ldots, x_n\}$. Now V is also a right $K[G]$-module and as such it is faithful. Since right and left multiplication commute as operators on V, it follows tht $K[G]$ is a set of E-linear transformations on an n-dimensional free E-module V. This $K[G] \subseteq E_n = E \otimes_K K_n$. By Theorem 4.6, K_n satisfies s_{2n}. Furthermore s_{2n} is multilinear and E is a commutative ring, so clearly $E_n = E \otimes_K K_n$ also satisfies s_{2n}. Since $K[G] \subseteq E_n$, the result follows.

Suppose A is an abelian subgroup of G with $[G:A] < \infty$. Then every element of A has only a finite number of conjugates in G and thus $\Delta(G) \supseteq A$ and $[G:\Delta] < \infty$. Therefore according to the observation of Ref. [*50*], a first step in finding a converse to Theorem 5.1 is to show that $[G:\Delta] < \infty$. That is the goal of the remainder of this section.

Lemma 5.2. Let G be a group and suppose that G can be written as $G = \bigcup H_i x_{ij}$, a finite union of cosets. Then $G = \bigcup' H_i x_{ij}$ where the union is restricted to those H_i with $[G:H_i] < \infty$.

Proof. Let $\mathscr{S} = \{i \mid [G:G_i] < \infty\}$ and let $\mathscr{T} = \{i \mid [G:H_i] = \infty\}$. By Lemma 1.2, $\mathscr{S} \neq \varnothing$. Let $W = \bigcap_{i\in\mathscr{S}} H_i$. Then $[G:W] < \infty$ by Lemma 1.1 and each coset $H_i x_{ij}$ with $i \in \mathscr{S}$ is a finite union of cosets of W. Thus

$$\bigcup{}' H_i x_{ij} = \bigcup_{i\in\mathscr{S}} H_i x_{ij} = \bigcup W y_k$$

a finite union of cosets of W. If $G \neq \bigcup' H_i x_{ij}$, then $G \neq \bigcup W y_k$ and some coset Wy is missing. Then

$$Wy \subseteq \left(\bigcup W y_k\right) \cup \left(\bigcup_{i\in\mathscr{T}} H_i x_{ij}\right)$$

and since $Wy \cap Wy_k$ is empty we have $Wy \subseteq \bigcup_{i\in\mathscr{T}} H_i x_{ij}$. Thus all cosets of W are contained in finite unions of cosets of those H_i with $i \in \mathscr{T}$. Since $[G:W] < \infty$, this yields a representation of G as a finite union of cosets of those H_i with $i \in \mathscr{T}$. This contradicts Lemma 1.2 and thus $G = \bigcup' H_i x_{ij}$.

Lemma 5.3. Let $G \neq \bigcup H_m g_{mn}$, a finite union of cosets. Let $\alpha_1, \alpha_2, \ldots, \alpha_s, \beta_1, \beta_2, \ldots, \beta_s \in K[G]$ and suppose that for all $x \in G - \bigcup H_m g_{mn}$ we have

$$\alpha_1 x \beta_1 + \alpha_2 x \beta_2 + \cdots + \alpha_s x \beta_s = 0.$$

Then there exists $y \in G$ with

$$\theta(\alpha_1)^y \beta_1 + \theta(\alpha_2)^y \beta_2 + \cdots + \theta(\alpha_s)^y \beta_s = 0.$$

Proof. Let W be the intersection of the centralizers of all the elements in $\operatorname{Supp} \theta(\alpha_i)$ for all $i = 1, 2, \ldots, s$. By Lemma 1.1, $[G:W] = t < \infty$. Clearly if $x \in W$, then x centralizes $\theta(\alpha_1), \theta(\alpha_2), \ldots, \theta(\alpha_s)$. Let $\{u_i\}$ be a set of right coset representatives for W in G. Let us suppose by way of contradiction that for $i = 1, 2, \ldots, t$,

$$\gamma_i = \theta(\alpha_1)^{u_i}\beta_1 + \theta(\alpha_2)^{u_i}\beta_2 + \cdots + \theta(\alpha_s)^{u_i}\beta_s \neq 0$$

and let $v_i \in \operatorname{Supp} \gamma_i$.

Write $\alpha_j = \theta(\alpha_j) + \alpha_j'$ where $\operatorname{Supp} \alpha_j' \cap \Delta = \varnothing$ and then write the finite sums

$$\alpha_j' = \sum a_{jk} y_k \qquad y_k \notin \Delta$$
$$\beta_j = \sum b_{jk} z_k$$

with $a_{jk}, b_{jk} \in K$ and $y_k, z_k \in G$. If y_j is conjugate to some $v_i z_k^{-1}$ in G, choose $h_{ijk} \in G$ with $h_{ijk}^{-1} y_j h_{ijk} = v_i z_k^{-1}$.

Let $x \in G$ and suppose that $x \notin \bigcup H_m g_{mn}$. Then we must have

$$\begin{aligned} 0 &= x^{-1}\alpha_1 x\beta_1 + x^{-1}\alpha_2 x\beta_2 + \cdots + x^{-1}\alpha_s x\beta_s \\ &= [\theta(\alpha_1)^x\beta_1 + \theta(\alpha_2)^x\beta_2 + \cdots + \theta(\alpha_s)^x\beta_s] \\ &\quad + [\alpha_1'^x\beta_1 + \alpha_2'^x\beta_2 + \cdots + \alpha_s'^x\beta_s]. \end{aligned}$$

Since $\{u_i\}$ is a full set of coset representatives for W in G, we have $x \in Wu_i$ for some i. Since W centralizes $\theta(\alpha_1), \theta(\alpha_2), \ldots, \theta(\alpha_s)$, the first expression above is equal to γ_i. Hence

$$0 = \gamma_i + [\alpha_1'^x\beta_1 + \alpha_2'^x\beta_2 + \cdots + \alpha_s'^x\beta_s].$$

Now v_i occurs in the support of γ_i and so this element must be canceled by something from the second term. Thus there exists y_j, z_k with $v_i = y_j^x z_k$ or

$$x^{-1} y_j x = v_i z_k^{-1} = h_{ijk}^{-1} y_j h_{ijk}.$$

Thus $x \in \mathbf{C}_G(y_j)h_{ijk}$. We have therefore shown that

$$G = \left(\bigcup H_m g_{mn}\right) \cup \left(\bigcup \mathbf{C}_G(y_j)h_{ijk}\right),$$

a finite union of cosets. Now $y_j \notin \Delta$ so $[G : \mathbf{C}_G(y_j)] = \infty$. Since by Lemma 5.2 we can delete subgroups of infinite index from the above union, we have $G = \bigcup H_m g_{mn}$, a contradiction. The lemma is proved.

Let $K[\zeta_1, \zeta_2, \ldots]$ be the polynomial ring over K in the noncommuting indeterminates $\zeta_1, \zeta_2, \ldots$. A linear monomial is an element $\mu \in K[\zeta_1, \zeta_2, \ldots]$ of the form $\mu = \zeta_{i_1}\zeta_{i_2}\cdots\zeta_{i_r}$ with all i_j distinct and with $r \geq 1$. Thus μ is linear in each variable.

Lemma 5.4. The number of linear monomials in $K[\zeta_1, \zeta_2, \ldots, \zeta_m]$ is less than or equal to $(m + 1)!$.

Proof. The number of linear monomials in $K[\zeta_1, \zeta_2, \ldots, \zeta_m]$ of degree m is of course $m!$. Now any other linear monomial is clearly just an initial segment of one of these. This yields the bound $m \cdot m! \leq (m+1)!$.

Theorem 5.5 (Passman [*44*]). Let $K[G]$ satisfy a nontrivial polynomial identity of degree n. Then $[G:\Delta] \leq n!$.

Proof. We assume by way of contradiction that $[G:\Delta] > n!$. By Lemma 4.1 we may assume that $K[G]$ satisfies the polynomial identity

$$f(\zeta_1, \zeta_2, \ldots, \zeta_n) = \zeta_1\zeta_1 \cdots \zeta_n + \sum_{\substack{\sigma \in S_n \\ \sigma \neq 1}} a_\sigma \, \zeta_{\sigma(1)} \, \zeta_{\sigma(2)} \cdots \zeta_{\sigma(n)}$$

so that clearly $n > 1$. For $j = 1, 2, \ldots, n$, define $f_j \in K[\zeta_j, \zeta_{j+1}, \cdots, \zeta_n]$ by

$$f = \zeta_1 \, \zeta_2 \cdots \zeta_{j-1} f_j + \text{terms not starting with } \zeta_1 \, \zeta_2 \cdots \zeta_{j-1}.$$

Then clearly $f_1 = f$, $f_n = \zeta_n$, and f_j is a homogeneous multilinear polynomial of degree $n - j + 1$. In particular, for all j, ζ_j occurs in each monomial of f_j. We clearly have

$$f_j = \zeta_j f_{j+1} + \text{terms not starting with } \zeta_j.$$

For $j = 2, 3, \ldots, n$, let $\mathscr{M}_j$ denote the set of all linear monomials in $K[\zeta_j, \zeta_{j+1}, \ldots, \zeta_n]$ and let $\mathscr{M}_1$ be empty. Then by Lemma 5.4 we have for all j, $|\mathscr{M}_j| \leq |\mathscr{M}_2| \leq n!$. We show now by induction on $j = 1, 2, \ldots, n$ that for any $x_j, x_{j+1}, \ldots, x_n \in G$, then either $f_j(x_j, x_{j+1}, \ldots, x_n) = 0$ or $\mu(x_j, x_{j+1}, \ldots, x_n) \in \Delta$ for some $\mu \in \mathscr{M}_j$. Since $f = f_1$ is a polynomial identity satisfied by $K[G]$, the result for $j = 1$ is clear.

Suppose the result holds for some $j < n$. Fix $x_{j+1}, x_{j+2}, \ldots, x_n \in G$ and let $x \in G$ play the role of the jth variable. Let $\mu \in \mathscr{M}_{j+1}$. If $\mu(x_{j+1}, x_{j+2}, \ldots, x_n) \in \Delta$, we are done. Thus we may assume that $\mu(x_{j+1}, x_{j+2}, \ldots, x_n) \notin \Delta$ for all $\mu \in \mathscr{M}_{j+1}$. Set $\mathscr{M}_j - \mathscr{M}_{j+1} = \mathscr{T}_j$.

Now let $\mu \in \mathscr{T}_j$ so that μ involves the variable ζ_j. Write $\mu = \mu' \zeta_j \mu''$ where μ' and μ'' are monomials in $K[\zeta_{j+1}, \zeta_{j+2}, \ldots, \zeta_n]$. Then $\mu(x, x_{j+1}, x_{j+2}, \ldots, x_n) \in \Delta$ if and only if

$$x \in \mu'(x_{j+1}, \ldots, x_n)^{-1} \, \Delta \mu''(x_{j+1}, \ldots, x_n)^{-1} = \Delta h_\mu,$$

a fixed coset of Δ, since μ' and μ'' do not involve ζ_j and since Δ is normal in G. Thus it follows that for all $x \in G - \bigcup_{\mu \in \mathscr{T}_j} \Delta h_\mu$, we have $\mu(x, x_{j+1}, \ldots, x_n) \notin \Delta$ for all $\mu \in \mathscr{M}_j$, since $\mathscr{M}_j \subseteq \mathscr{M}_{j+1} \cup \mathscr{T}_j$. Since the inductive result holds for j, we conclude that for all $x \in G - \bigcup_{\mu \in \mathscr{T}_j} \Delta h_\mu$ we have $f_j(x, x_{j+1}, \ldots, x_n) = 0$. Note that $|\mathscr{T}_j| \leq |\mathscr{M}_j| \leq n!$ and $[G:\Delta] > n!$ by assumption, so $G - \bigcup_{\mu \in \mathscr{T}_j} \Delta h_\mu$ is nonempty.

Write

$$f_j(\zeta_j, \zeta_{j+1}, \ldots, \zeta_n) = \zeta_j f_{j+1} + \Sigma_r \eta_r \zeta_j \eta_r'$$

where $\eta_r, \eta_r' \in K[\zeta_{j+1}, \zeta_{j+2}, \ldots, \zeta_n]$ and η_r is a linear monomial. Hence $\eta_r \in \mathscr{M}_{j+1}$. Now by the above we have

$$0 = 1 \cdot x \cdot f_{j+1}(x_{j+1}, \ldots, x_n)$$
$$+ \Sigma_r \eta_r(x_{j+1}, \ldots, x_n) \cdot x \cdot \eta_r'(x_{j+1}, \ldots, x_n)$$

for all $x \in G - \bigcup_{\mu \in \mathscr{T}_j} \Delta h_\mu \neq \varnothing$. Hence by Lemma 5.3 there exists $y \in G$ with

$$0 = \theta(1)^y f_{j+1}(x_{j+1}, \ldots, x_n)$$
$$+ \Sigma_r \theta(\eta_r(x_{j+1}, \ldots, x_n))^y \eta_r'(x_{j+1}, \ldots, x_n).$$

Clearly $\theta(1)^y = 1$. Also $\eta_r(x_{j+1}, \ldots, x_n) \in G - \Delta$ since $\eta_r \in \mathscr{M}_{j+1}$ and hence $\theta(\eta_r(x_{j+1}, \ldots, x_n)) = 0$. Thus

$$0 = 1 \cdot f_{j+1}(x_{j+1}, \ldots, x_n) = f_{j+1}(x_{j+1}, \ldots, x_n)$$

and the induction step is proved.

In particular, the inductive result holds for $j = n$. Here $f_n(\zeta_n) = \zeta_n$ and $\mathscr{M}_n = \{\zeta_n\}$. Thus we conclude that for all $x \in G$ either $x = 0$ or $x \in \Delta$, a contradiction since $G \neq \Delta$. Therefore the assumption that $[G:\Delta] > n!$ is false and the theorem is proved.

§6. PRIME POLYNOMIAL IDENTITY RINGS

Lemma 6.1. Let G be a finitely generated group and let H be a subgroup of finite index. Then H is finitely generated.

Proof. By adding inverses if necessary, we can assume that G is generated by $x_1, x_2, \ldots, x_t$ as a semigroup. Let $y_1, y_2, \ldots, y_n$ be a set of right coset representatives for H in G. For each i, j, Hy_ix_j is a coset of H, say $Hy_ix_j = Hy_{i'}$. Then there exists $h_{ij} \in H$ with $y_ix_j = h_{ij}y_{i'}$. Let $\bar{H}$ be the subgroup of H generated by $\{h_{ij}\}$ and set $W = \bigcup \bar{H}y_i$. Since $h_{ij} \in \bar{H}$, we have $(\bar{H}y_i)x_j = \bar{H}h_{ij}y_{i'} = \bar{H}y_{i'} \subseteq W$ and hence $Wx_j = W$. Thus since the x_j generate G as a semigroup, we have $WG = W$ and hence clearly $W = G$. This yields easily $\bar{H} = H$ and the result follows.

Corollary 6.2. Let G be a finitely generated group and suppose that $K[G]$ satisfies a polynomial identity. Then G has a normal abelian subgroup of finite index.

Proof. By Theorem 5.5, $[G:\Delta] < \infty$ and hence by the previous lemma Δ is finitely generated. Thus by Lemma 2.2, $[\Delta:\mathbf{Z}(\Delta)] < \infty$ so $\mathbf{Z}(\Delta)$ is an abelian subgroup of G of finite index. Since $\mathbf{Z}(\Delta)$ is characteristic in Δ, it is normal in G.

We will see later by examples that even if we know the degree of the polynomial identity we cannot, in general, bound the index of the abelian subgroup in the above. Furthermore if G is not finitely generated, then G need not even have an abelian subgroup of finite index.

We now discuss a few results of a more general ring theoretic nature. Let R be a primitive ring, that is, R is a ring with a faithful irreducible right module V. Then by Schur's lemma, the commuting ring $D = \mathrm{Hom}_R(V, V)$ is a division ring, V is a left D-module, and R is a ring of D-linear transformations of V. Furthermore by the Density theorem [22], R is in fact a dense set of D-linear transformations on V. This means that given any finite number $v_1, v_2, \ldots, v_n$ of D-linearly independent elements of V and any arbitrary elements $w_1, w_2, \ldots, w_n$ of V, then there exists $\alpha \in R$ with $v_i\alpha = w_i$. Clearly if R is an algebra over a field K, then so also is D. The above facts are assumed to be known and will not be proved here.

Lemma 6.3. Let D be a division ring with center Z and let E be a maximal subfield of D. Then $E \otimes_Z D$ has a faithful irreducible module with commuting ring $E \otimes 1 \simeq E$.

Proof. We let $E \otimes_Z D$ act on D by

$$x(\textstyle\sum e_i \otimes d_i) = \sum e_i x d_i$$

where $e_i \in E$, $d_i \in D$, and $x \in D$. It is easy to see that this is a well-defined action and therefore that D is an $E \otimes_Z D$ module.

We show first that $E \otimes D$ acts faithfully. To do this it clearly suffices to show that if $e_1, e_2, \ldots, e_n \in E$ are linearly independent over Z and if $\sum e_i x d_i = 0$ for all $x \in D$, then $d_1 = d_2 = \cdots = d_n = 0$. We proceed by induction on n, the case $n = 1$ being obvious. Assume the result for $n - 1$ and let $\{e_i\}$ and $\{d_i\}$ be as above. Suppose by way of contradiction that some $d_i \neq 0$, say $d_n \neq 0$. Then multiplying $\sum e_i x d_i$ on the right by d_n^{-1} shows that we may assume that $d_n = 1$. Let $d \in D$. Then by subtracting the identities

$$\left(\sum_1^n e_i x d_i\right) d = 0 \qquad \text{and} \qquad \sum_1^n e_i (xd) d_i = 0$$

we obtain $\sum_1^{n-1} e_i x(d_i d - d d_i) = 0$. By induction we have $d_i d = d d_i$ for all $d \in D$ so $d_i \in Z$ for all i. Thus evaluating $\sum e_i x d_i = 0$ at $x = 1$ yields a nontrivial linear dependence of $\{e_i\}$ over Z, a contradiction, and $E \otimes D$ acts faithfully.

Let $x \in D$, $x \neq 0$. Since $xD = D$, we have $x(E \otimes D) = D$ and D is an irreducible $E \otimes D$ module. Finally let $\alpha \in \mathrm{Hom}_{E\otimes D}(D, D)$, the commuting ring. Then for $x \in D$, we have $\alpha(x) = \alpha(1 \cdot x) = \alpha(1)x$ since α commutes with right multiplication by D. If $e \in E$, we have in addition $e\alpha(1) = \alpha(e \cdot 1) = \alpha(1 \cdot e) = \alpha(1)e$ since α commutes with left multiplication by E. Thus $\alpha(1) \in D$ centralizes E and since E is a maximal subfield of D this implies that $\alpha(1) \in E$. Thus $\alpha = \alpha(1) \otimes 1$ in its action on D and the result follows.

Theorem 6.4 (Kaplansky [*25*]). Let R be a primitive algebra over a field K and suppose that R satisfies a polynomial identity of degree n. Then

(i) $R \simeq D_t$, the ring of $t \times t$ matrices over a division algebra D.
(ii) If Z is the center of D, then $\dim_Z R = m^2$ where m is an integer satisfying $1 \leq m \leq [n/2]$.
(iii) $R \subseteq E_m$, the ring of $m \times m$ matrices over some field E containing K.

Proof. By Lemma 4.1, R satisfies a multilinear polynomial f of degree n.

Let V be the given faithful irreducible R module and let D be the commuting ring. If $\dim_D V = \infty$, then it follows easily from the Density theorem that, for each positive integer s, R has a subalgebra S with K_s as a homomorphic image. Thus K_s also satisfies f and this is a contradiction for $2s > n$ by Lemma 4.2. Thus $\dim_D V = t < \infty$ and by the Density theorem again $R \simeq D_t$ so (i) follows.

Let Z be the center of D and let E be a maximal subfield whose existence is guaranteed by Zorn's lemma. By the previous lemma, $E \otimes_Z D$ has a faithful irreducible representation with commuting ring $E \otimes 1 \simeq E$. Since $R \simeq D_t$ contains an isomorphic copy of D, we see that D satisfies f and since f is multilinear and E is commutative, $E \otimes D$ also satisfies f. Thus the argument of the preceding paragraph applies to $E \otimes D$ and $E \otimes D \simeq E_{t'}$ for some integer t'. Since $R = D \otimes_Z Z_t$ this yields easily $E \otimes_Z R = E \otimes_Z Z_m = E_m$ where $m = tt'$. Again $E \otimes_Z R$ satisfies f so by Lemma 4.2 we have $m \leq [n/2]$. Now

$$\dim_Z R = \dim_E E \otimes_Z R = \dim_E E_m = m^2$$

so (ii) follows. Finally $R \subseteq E \otimes_Z R = E_m$ so (iii) follows.

We now return to the study of group rings. Under certain circumstances the bound on $[G:\Delta]$ in Theorem 5.5 can be improved. The following result can be found in [*50*]. We obtain it here as a corollary of Theorem 5.5 but chronologically it came first and in fact motivated that theorem. The proof below retains the basic flavor of the original, namely the formation of a suitable ring of quotients, but it does not require the use of certain ring theoretic machinery. Amazingly enough we apply some elementary Galois theory. In the next section, we will give the original proof.

Theorem 6.5 (Smith [*50*]). Let $K[G]$ be prime and suppose that $K[G]$ satisfies a polynomial identity of degree n. Then $\Delta(G)$ is torsion free abelian and $[G:\Delta] \leq n/2$.

Proof. By Theorem 2.5, Δ is torsion free abelian and by Theorem 5.5, $[G:\Delta] = k < \infty$. Hence by Lemma 2.4, no nonzero element of $K[\Delta]$ is a zero divisor in $K[G]$ and in particular $K[\Delta]$ is an integral domain. Set $\bar{G} = G/\Delta$. Then $\bar{G}$ acts faithfully by conjugation on Δ since if $x \in G$ and x centralizes Δ, then $[G:\mathbf{C}_G(x)] < \infty$ and $x \in \Delta$. Thus $\bar{G}$ acts faithfully by conjugation as ring automorphisms on $K[\Delta]$. Let $x_1, x_2, \ldots, x_k$ be a complete set of coset representatives for Δ in G with $x_1 = 1$.

Let Z denote the center of $K[G]$. As we observed earlier, $Z \subseteq K[\Delta]$ and thus no nonzero element of Z is a zero divisor in $K[G]$. Since Z is central it is then trivial to form the ring of quotients $Z^{-1}K[G]$. This is the set of all formal fractions $\eta^{-1}\alpha$ with $\eta \in Z - \{0\}$, $\alpha \in K[G]$, and with the usual identifications made.

Let $L = Z^{-1}K[\Delta] \subseteq Z^{-1}K[G]$ and let $F = Z^{-1}Z \subseteq L$. Clearly F is a field and L is an integral domain. Suppose $\alpha \in K[\Delta]$, $\alpha \neq 0$. Then $\alpha(\alpha^{x_2}\alpha^{x_3}\cdots\alpha^{x_k}) \in Z - \{0\}$ since $K[\Delta]$ is commutative. Thus α is invertible in L and L is a field. Now $\bar{G}$ acts on L and in fact we see that $\bar{G}$ is a group of field automorphisms of L with fixed field precisely F. The latter follows since if $\eta^{-1}\alpha \in L$ is fixed by all elements of $\bar{G}$, then $\alpha \in Z$ and $\eta^{-1}\alpha \in F$. Thus by Galois theory [*7*, Theorem 14], $(L:F) = |\bar{G}| = k$. Since $K[G]$ is free over $K[\Delta]$ of rank k, this shows that $E = Z^{-1}K[G]$ is a finite dimensional algebra over F and $\dim_F E = k^2$.

We observe now that E is prime. Suppose A and B are ideals of E with $AB = 0$. Then $A \cap K[G]$ and $B \cap K[G]$ are ideals in $K[G]$ with $(A \cap K[G])(B \cap K[G]) = 0$. Since $K[G]$ is prime, we conclude that one of the two ideals is zero, say $A \cap K[G] = 0$. Let $\eta^{-1}\alpha \in A$. Then $\alpha = \eta(\eta^{-1}\alpha) \in A \cap K[G]$ so $\alpha = 0$ and $\eta^{-1}\alpha = 0$. Thus $A = 0$ and E is prime. Since E is a finite dimensional algebra, we conclude first that E is semisimple and then by the Wedderburn theorems that E is simple. Thus E is a primitive ring.

Now by Lemma 4.1, we can assume that $K[G]$ satisfies a multilinear polynomial identity of degree n. Since Z is central, it follows that E also satisfies this identity viewed as a polynomial over F. Since F is clearly the center of E it follows from Theorem 6.4 that $k^2 = \dim_F E \leq [n/2]^2$. Thus $[G:\Delta] = k \leq [n/2]$.

Lemma 6.6 Suppose $K[G]$ satisfies a polynomial identity f of degree n. Let H be a subgroup of G. Then $K[H]$ also satisfies f. Furthermore if H is normal in G, then $K[G/H]$ satisfies f.

Proof. The first statement is clear since $K[H] \subseteq K[G]$. Suppose H is normal in G. Then the homomorphism $G \to G/H$ induces an epimorphism $K[G] \to K[G/H]$, so the second result follows.

Corollary 6.7. Suppose G is finitely generated and $K[G]$ satisfies a polynomial identity of degree n. Then $[G:\Delta] \leq n/2$.

Proof. By Theorem 5.5, $[G:\Delta] < \infty$ and hence by Lemma 6.1, Δ is finitely generated. Thus by Lemma 2.2, Δ' is finite. Since Δ/Δ' is a finitely generated abelian group and Δ' is finite, we conclude that H, the set of all elements of finite order in Δ, is in fact a finite subgroup of Δ. Clearly H is normal in G.

Set $\bar{G} = G/H$ and $\bar{\Delta} = \Delta/H$ so that $\bar{\Delta} \subseteq \Delta(\bar{G})$. On the other hand, suppose $\bar{x} = Hx \in \Delta(\bar{G})$. Then the conjugates of x are contained in only finitely many cosets of H and since H is finite, $x \in \Delta$. Thus $\bar{\Delta} = \Delta(\bar{G})$. Since $\bar{\Delta}$ is clearly torsion free abelian, we see that $K[\bar{G}]$ is prime by Theorem 2.5. Furthermore by Lemma 6.6, $K[\bar{G}]$ satisfies a polynomial identity of degree n. Hence by Theorem 6.5, $[\bar{G}:\bar{\Delta}] \leq n/2$ and since $[G:\Delta] = [\bar{G}:\bar{\Delta}]$, the result follows.

§7. ORDERS

Let R be a ring. A ring $Q(R) \supseteq R$ is said to be a complete ring of quotients of R if

(i) Every element of R which is not a zero divisor in R is invertible in $Q(R)$.

(ii) Every element of $Q(R)$ is of the form $\alpha^{-1}\beta$ where $\alpha, \beta, \in R$ and α is not a zero divisor in R.

If $Q(R)$ is a complete ring of quotients of R, then we say that R is an order in $Q(R)$.

In this section we offer an interesting result of Martha Smith on the center of $Q(K[G])$, if the latter exists, and we sketch the original proof of Theorem 6.5.

Lemma 7.1. Let $K[G]$ be a semiprime group ring with center C. Let H be a finitely generated normal subgroup of G contained in $\Delta(G)$ and let I be a right ideal in $K[H]$.

(i) Then either there exists $\beta_1 \in K[H]$, $\beta_1 \neq 0$ with $\beta_1 I = 0$, or there exists $\beta_2 \in I \cap C$ such that β_2 is not a zero divisor in $K[G]$.

(ii) Suppose further that $x^{-1}Ix = I$ for all $x \in G$ and that $I \neq 0$. Then $I \cap C \neq 0$.

Proof. By Lemma 2.2, $[H:\mathbf{Z}(H)] < \infty$ and certainly $\mathbf{Z}(H)$ is normal in G. Since H is finitely generated, it follows that $\mathbf{Z}(H)$ is a finitely generated abelian group by Lemma 6.1. Thus $\mathbf{Z}(H) = T \times B$ where T is a finitely generated torsion free abelian group and B is finite of order n. If $A = \{x^n \mid x \in \mathbf{Z}(H)\}$, then clearly A is normal in G and A is a torsion free central subgroup of H of finite index.

Set $Z = K[A]$ so that Z is a central subalgebra of $K[H]$. Moreover by Lemma 2.4 no nonzero element of Z is a zero divisor in $K[G]$ since A is torsion free abelian. It is then trivial to form the ring of quotients $E = Z^{-1}K[H]$. This is the set of all formal fractions $\eta^{-1}\alpha$ with $\eta \in Z$, $\eta \neq 0$, $\alpha \in K[H]$, and with the usual identifications made. If $F = Z^{-1}K[A]$, then F is certainly a central subfield of E and since $[H:A] < \infty$, E is a finite dimensional F-algebra.

We observe now that E is semiprime. Since $K[G]$ is semiprime, it follows from Theorem 3.7 that $H \subseteq \Delta(G)$ has no elements of order p in case K has characteristic p. Thus by Theorems 3.3 and 3.7, $K[H]$ is semiprime. Suppose L is an ideal of E with square zero. Then $L \cap K[H]$ is an ideal of $K[H]$ with square zero and hence $L \cap K[H] = 0$ since $K[H]$ is semiprime. Now if $\eta^{-1}\alpha \in L$, then $\alpha = \eta(\eta^{-1}\alpha) \in L \cap K[H]$ so $\alpha = 0$. Thus $L = 0$ and E is semiprime. Since E is a finite dimensional algebra this therefore implies that E is a semisimple Artinian ring.

We will need the following. Suppose $\alpha \in K[A]$, $\alpha \neq 0$. Since $A \subseteq \Delta(G)$, α has only finitely many distinct conjugates $\alpha_1 = \alpha, \alpha_2, \ldots, \alpha_t$ under the action of G. Hence since A is abelian and normal in G, the product $\gamma = \alpha_1\alpha_2 \cdots \alpha_t$ is central in $K[G]$. Moreover as we observed above, each α_i is not a zero divisor in $K[G]$. We have therefore shown that there exists $\tilde{\alpha} \in K[A]$ such that $\alpha\tilde{\alpha}$ is a nonzero element of C which is not a zero divisor in $K[G]$.

Now $Z^{-1}I$ is a right ideal in E since Z is central in $K[H]$. Thus by the Wedderburn theorems there exists an idempotent $e \in Z^{-1}I$ with $Z^{-1}I = eE$.

(i) Suppose first that $e = 1$ and write $e = \eta^{-1}\alpha$ with $\alpha \in I$, $\eta \in Z$, $\eta \neq 0$. Then $\alpha = \eta \in I \cap K[A]$ and $\alpha \neq 0$. By the above, $\beta_2 = \alpha\tilde{\alpha}$ is a nonzero element of $I \cap C$ which is not a zero divisor in $K[G]$. Now suppose that $e \neq 1$. Then $1 - e \neq 0$ and write $1 - e = \mu^{-1}\beta$ with $\beta \in K[H]$, $\beta \neq 0$, $\mu \in Z$, $\mu \neq 0$. Since $I \subseteq Z^{-1}I$, we then have clearly

$$\beta I \subseteq \mu(1 - e) \cdot eE = 0$$

and (i) follows with $\beta_1 = \beta$.

(ii) We suppose further that $x^{-1}Ix = I$ for all $x \in G$. If $x \in H$, then since I is a right ideal in $K[H]$ we have $xI = Ix = I$ so I is in fact a two sided ideal in $K[H]$. Thus $Z^{-1}I$ is a two sided ideal in E and by the Wedderburn theorems we may take e above to to be central in E so that e is in fact the unique identity element of $Z^{-1}I$. Now H and A are normal in G so G acts by conjugation on $K[H]$ and on $Z = K[A]$ and hence G acts on $E = Z^{-1}K[H]$. By assumption, G leaves $Z^{-1}I$ setwise invariant and thus G must fix e, the identity of $Z^{-1}I$. Write $e = \eta^{-1}\alpha$ with $\alpha \in I$, $\eta \in Z$, $\eta \neq 0$. Since e is central in E, we then have also

$$e = (\tilde{\eta})^{-1}(\eta^{-1}\alpha)\tilde{\eta} = (\eta\tilde{\eta})^{-1} \cdot (\alpha\tilde{\eta}).$$

Now $\eta\tilde{\eta} \in C$ and e is fixed under conjugation by G so this clearly yields $\alpha\tilde{\eta} \in I \cap C$. Since $I \neq 0$, we have $e \neq 0$ so $\alpha\tilde{\eta} \neq 0$ and the result follows.

Lemma 7.2. Let $K[G]$ be a semiprime group ring with center C. If $\alpha \in C$, then α is a zero divisor in $K[G]$ if and only if it is a zero divisor in C.

Proof. If α is a zero divisor in C, then certainly it is a zero divisor in $K[G]$. We consider the converse. Let α be a zero divisor in $K[G]$ and let $\beta \in K[G]$, $\beta \neq 0$ with $\alpha\beta = 0$. Let H be the subgroup of G generated by the support of α. Since $\alpha \in C$, we conclude that H is a finitely generated normal subgroup of G with $H \subseteq \Delta(G)$. Write $\beta = \sum \beta_i x_i$ with $\beta_i \in K[H]$ and with the x_i in distinct cosets of H in G. It then follows that $\alpha\beta_i = 0$ for all i.

Let $I = \{\gamma \in K[H] \mid \alpha\gamma = 0\}$. Since $\beta_i \in I$ for all i and $\beta \neq 0$, we conclude that $I \neq 0$. Clearly I is a right ideal in $K[H]$. Moreover since α is central in $K[G]$ and H is normal in G, we have for $x \in G$, $x^{-1}Ix \subseteq K[H]$ and

$$\alpha x^{-1}Ix = x^{-1}(\alpha I)x = 0.$$

Thus $x^{-1}Ix \subseteq I$. This yields $x^{-1}Ix = I$ and by Lemma 7.1(ii) we have $I \cap C \neq 0$. Thus α is a zero divisor in C and the result follows.

Lemma 7.3. Let $K[G]$ be a semiprime group ring and let $\alpha \in K[G]$. If α is not a right divisor of zero, then there exists $\gamma \in K[G]$ such that $\theta(\alpha\gamma)$ is central and not a zero divisor in $K[G]$.

Proof. Write $\alpha = \sum_1^n \alpha_i x_i$ with $\alpha_i \in K[\Delta]$ and with the x_i in distinct cosets of Δ. Let H be the subgroup of G generated by the elements in the support of all α_i and their finitely many conjugates. Then H is a finitely generated normal subgroup of G and $H \subseteq \Delta(G)$. Let I be the right ideal in $K[H]$ generated by the α_i. Thus

$$I - \sum_1^n \alpha_i K[H].$$

Suppose there exists $\beta_1 \in K[H]$, $\beta_1 \neq 0$ such that $\beta_1 I = 0$. Then clearly $\beta_1\alpha = 0$ and α is a right divisor of zero, a contradiction. Thus by Lemma 7.1(i) we conclude that there exists $\beta = \beta_2 \in I$ such that β is a central element in $K[G]$ which is not a zero divisor.

Since $\beta \in I$, we have $\beta = \sum_1^n \alpha_i\gamma_i$ with $\gamma_i \in K[H]$. Set

$$\gamma = \sum_1^n x_i^{-1}\gamma_i \in K[G].$$

Then $\alpha\gamma = \sum_{i,j} \alpha_i x_i x_j^{-1}\gamma_j$. If $i \neq j$, then clearly $\operatorname{Supp}(\alpha_i x_i x_j^{-1}\gamma_j) \cap \Delta = \varnothing$ since Δ is normal in G and $x_i x_j^{-1} \notin \Delta$. Thus

$$\theta(\alpha\gamma) = \sum_1^n \alpha_i\gamma_i = \beta$$

and the lemma is proved.

We now come to the main theorem of this section.

Theorem 7.4 (Smith [*50*]). Let $K[G]$ be a semiprime group ring which is an order in a ring Q. Then the center of $K[G]$ is an order in the center of Q.

Proof. Let C denote the center of $K[G]$ and let Z denote the center of Q so that clearly $Z \supseteq C$. Let $\alpha \in C$ be an element which is not a zero divisor in C. Then by Lemma 7.2, α is not a zero divisor in $K[G]$ so α is invertible in Q. Clearly $\alpha^{-1} \in Z$.

Now let $\rho \in Z \subseteq Q$. Then $\rho = \alpha^{-1}\beta$ where $\alpha, \beta \in K[G]$ and α is not a zero divisor in $K[G]$. Thus for all $\omega \in Q$, we have $\omega\alpha^{-1}\beta = \alpha^{-1}\beta\omega$ so $\alpha\omega\alpha^{-1}\beta = \beta\omega$. Now by Lemma 7.3 there exists $\gamma \in K[G]$ such that

$\theta(\alpha\gamma) \in C$ is not a zero divisor in $K[G]$. Set $\omega = \gamma x \alpha$ in the above. Then for all $x \in G$ we have

$$(\alpha\gamma)x\beta = (\beta\gamma)x\alpha$$

and Lemma 1.3 yields $\theta(\alpha\gamma)\beta = \theta(\beta\gamma)\alpha$. Set $\mathscr{E} = \theta(\alpha\gamma)$, $\eta = \theta(\beta\gamma)$. Since $\mathscr{E} \in C$ is not a zero divisor in $K[G]$, we have $\mathscr{E}^{-1}\eta \in Q$ and since $\alpha^{-1}\beta \in Z$ we obtain from the above

$$\begin{aligned}\mathscr{E}^{-1}\eta &= \beta\alpha^{-1} = \alpha(\alpha^{-1}\beta)\alpha^{-1}\\ &= \alpha^{-1}\beta = \rho.\end{aligned}$$

Finally $\mathscr{E} \in Z$, $\rho \in Z$ so $\eta = \mathscr{E}\rho \in Z \cap K[G] = C$ and $\rho = \mathscr{E}^{-1}\eta$ is a quotient of elements in C. The result follows.

We can now sketch the original proof due to Martha Smith of Theorem 6.5. To start with we quote the following.

Theorem (Posner [*47*]). Suppose that R is a prime ring satisfying a polynomial identity over its center. Then R can be embedded as an order in a full matrix ring $Q = D_n$ where D is a division algebra finite dimensional over its center.

See Refs. [*16*] and [*32*] for additional proofs.

Now suppose that $K[G]$ is prime and satisfies a polynomial identity and let $Q = Q(K[G])$ be the complete ring of quotients of $K[G]$ whose existence is guaranteed by the above result. We will show that $[G:\Delta] < \infty$ and in fact that $[G:\Delta] \leq \dim_Z Q$ where Z is the center of Q. Suppose that this is not the case and choose $t > \dim_Z Q$ elements $x_1, x_2, \ldots, x_t \in G$ in distinct cosets of Δ. Then $\{x_i\} \subseteq Q$ and since $t > \dim_Z Q$ it follows that there exist $\rho_1, \rho_2, \ldots, \rho_t \in Z$ not all zero with

$$\rho_1 x_1 + \rho_2 x_2 + \cdots + \rho_t x_t = 0.$$

By Theorem 7.4, the center C of $K[G]$ is an order in Z and hence there exists $\alpha_i, \beta_i \in C$ such that $\rho_i = \alpha_i^{-1}\beta_i$. Thus if $\alpha = \alpha_1 \alpha_2 \cdots \alpha_t$, then we have

$$\sum_1^t (\alpha\alpha_i^{-1}\beta_i)x_i = 0.$$

Now $\alpha\alpha_i^{-1}\beta_i \in C \subseteq K[\Delta]$ and the x_i are in distinct cosets of Δ in G so we conclude that $\alpha\alpha_i^{-1}\beta_i = 0$ for all i. Moreover since the α_i are not divisors of zero, this yields $0 = \alpha_i^{-1}\beta_i = \rho_i$, a contradiction. We have therefore shown that $[G:\Delta] < \infty$. A closer study of Q and a sharper version of Posner's theorem relating the degree of the original polynomial identity to

$\dim_Z Q$ yield the stronger result $[G:\Delta] \leq [n/2]$. We will not go into that here.

We close this section with a number of interesting results of a different nature.

Theorem 7.5 (Smith [*50*]). Let $K[G]$ be a prime group ring and let F be a fixed subfield of K. If I is a nonzero ideal in $K[G]$, then there exists $\alpha \in K[\Delta]$, $\beta \in F[G]$ such that α is central in $K[G]$ and $\alpha\beta \in I$, $\alpha\beta \neq 0$.

Proof. Since $K[G]$ is prime, Δ is a torsion free abelian group by Theorem 2.5 and hence by Lemma 2.4 no nonzero element of $K[\Delta]$ is a zero divisor in $K[G]$.

If $I \cap F[G] \neq 0$, take $\alpha = 1$, $\beta \in I \cap F[G]$ with $\beta \neq 0$. Thus we may suppose that $I \cap F[G] = 0$. Let $\{k_i\}$ be a basis for K/F. Then every element of $K[G]$ can be written as a finite sum $\sum \gamma_i k_i$ with $\gamma_i \in F[G]$. Now choose $\gamma \in I$, $\gamma \neq 0$ so that $\gamma = \sum \gamma_i k_i$ and with the minimal possible number of nonzero γ_i occurring. By reordering the k_i if necessary, we can assume that

$$\gamma = \sum_1^n \gamma_i k_i, \qquad \gamma_i \neq 0.$$

Also $n \geq 2$ since if $n = 1$, then $\gamma k_1^{-1} \in I \cap F[G]$. By replacing γ by $y\gamma$ for some $y \in G$ if necessary, we can clearly assume that $1 \in \operatorname{Supp} \gamma$ and say $1 \in \operatorname{Supp} \gamma_1$.

Let $x \in G$. Then $\gamma_1 x\gamma - \gamma x\gamma_1 \in I$ and

$$\gamma_1 x\gamma - \gamma x\gamma_1 = \sum_2^n (\gamma_1 x\gamma_i - \gamma_i x\gamma_1)k_i.$$

Thus by the minimality of n and the fact that the k_i are linearly independent over F, we conclude that

$$\gamma_1 x\gamma_i = \gamma_i x\gamma_1$$

for all $x \in G$ and for all i. Hence by Lemma 1.3 we have $\theta(\gamma_1)\gamma_i = \theta(\gamma_i)\gamma_1$. Now Δ is abelian so all $\theta(\gamma_i)$ commute and therefore

$$\theta(\gamma_1)\gamma = \sum_1^n \theta(\gamma_1)\gamma_i k_i = \left(\sum_1^n \theta(\gamma_i)k_i\right)\gamma_1.$$

Note that since $1 \in \operatorname{Supp} \gamma_1$, we have $\theta(\gamma_1) \neq 0$ and since $\theta(\gamma_1)$ is not a zero divisor in $K[G]$ we have $\theta(\gamma_1)\gamma \neq 0$.

Set $\beta = \gamma_1 \in F[G]$ and $\alpha_1 = \sum_1^n \theta(\gamma_i)k_i \in K[\Delta]$. Then $\alpha_1\beta = \theta(\gamma_1)\gamma \neq 0$ so $\alpha_1 \neq 0$. Let $\alpha = \alpha_1 \alpha_2 \cdots \alpha_t$ be the product of the finite number of distinct conjugates of α_1 under the action of G. Since $K[\Delta]$ is commutative it follows that α is central in $K[G]$ and we have

$$\alpha\beta = \alpha_1 \alpha_2 \cdots \alpha_t \beta = \alpha_2 \cdots \alpha_t \theta(\gamma_1)\gamma \in I.$$

Moreover each $\alpha_2, \alpha_3, \ldots, \alpha_t$ is not a zero divisor in $K[G]$ so $\alpha\beta \neq 0$ and the result follows.

Corollary 7.6. Let $K[G]$ be a prime group ring and let F be a fixed subfield of K. Let I be a nonzero ideal in $K[G]$.

(i) If I is a prime ideal, then either $I \cap F[G] \neq 0$ or $I \cap C \neq 0$ where C is the center of $K[G]$.

(ii) If $\Delta = \langle 1 \rangle$, then $I \cap F[G] \neq 0$.

Proof. We apply Theorem 7.5 and we use its notation.

(i) Since α is central in $K[G]$, we have

$$(K[G]\alpha K[G])(K[G]\beta K[G]) = (K[G]\alpha\beta K[G]) \subseteq I.$$

Thus since I is prime we must have either $\alpha \in I$ or $\beta \in I$.

(ii) Since $\alpha \neq 0$, $\alpha \in K[\Delta]$, and $\Delta = \langle 1 \rangle$, we see that $\alpha \in K$, $\alpha \neq 0$. Thus $\beta = \alpha^{-1}(\alpha\beta) \in I \cap F[G]$ and the result follows.

II

BOUNDED REPRESENTATION DEGREE

§8. IRREDUCIBLE REPRESENTATIONS

Let E be an algebra over a field K. An ideal P of E is said to be primitive if E/P is a primitive ring, that is, if E/P has a faithful irreducible right module V. The intersection of all primitive ideals of E is JE, the Jacobson radical of E. We say that E is semisimple if and only if $JE = 0$. Basic properties of JE will be discussed in the next chapter. Here we merely observe that E is semisimple if and only if it is a subdirect product of primitive rings.

A representation of E is a homomorphism of E into $\mathrm{End}(V)$, the ring of endomorphisms of some additive abelian group V. In this case, of course, V becomes an E-module. Conversely any E-module V corresponds to a representation of E. A representation of E is said to be irreducible if the associated E-module V is irreducible. Let ρ be an irreducible representation of E. Then $\rho(E)$ is a ring with a faithful irreducible representation and hence is a primitive ring. Thus the kernel of ρ is a primitive ideal in E. Conversely if P is a primitive ideal in E, then E/P has a faithful irreducible module V and the combined map $E \to E/P \to \mathrm{End}(V)$ yields an irreducible representation of E with kernel P.

Let ρ be an irreducible representation of the algebra E. Then ρ is said to be of finite degree if the algebra $\rho(E)$ satisfies a polynomial identity over K. In this case, Theorem 6.4 applies and if Z denotes the center of $\rho(E)$, then Z is a field and $\dim_Z \rho(E) = m^2$ for some positive integer m. We set the degree of ρ equal to this integer m.

Let n be a positive integer. We say E has r.b. n, representation bound n, if all irreducible representations ρ of E have finite degree less than or equal to n.

Theorem 8.1 (Kaplansky [*25*]). Let E be an algebra over K and let n be an integer. Consider the possible properties of E.

(i) E has r.b. $[n/2]$.

(ii) E satisfies a polynomial identity of degree n.

(iii) E satisfies the standard identity of degree n.

(iv) $E \subseteq R_{[n/2]}$, the ring of $[n/2] \times [n/2]$ matrices over a commutative algebra R.

Then (iv) $\Rightarrow$ (iii) $\Rightarrow$ (ii) $\Rightarrow$ (i). If in addition E is semisimple, then (i) $\Rightarrow$ (iv).

Proof. (iv) $\Rightarrow$ (iii) By Theorem 4.6 and Lemma 4.3(iv), $K_{[n/2]}$ satisfies s_n, the standard identity of degree n. Since R is commutative and s_n is multilinear, it follows that $R_{[n/2]}$ also satisfies s_n and hence so does E.

(iii) $\Rightarrow$ (ii). This is obvious.

(ii) $\Rightarrow$ (i). Let E satisfy f, a polynomial identity of degree n and let ρ be an irreducible representation of E. Then $\rho(E)$ also satisfies f and hence ρ is of finite degree. If Z denotes the center of $\rho(E)$, then by Theorem 6.4, $\dim_Z \rho(E) = m^2$ and $m \leq [n/2]$. Thus E has r.b. $[n/2]$.

(i) $\Rightarrow$ (iv). We assume here that E is semisimple and has r.b. $[n/2]$. Let ρ be an irreducible representation of E. Then by Theorem 6.4, $\rho(E) \subseteq (R\rho)_m$ where $R\rho$ is some field extension of K and m is the degree of ρ. Since $m \leq [n/2]$, we then have $\rho(E) \subseteq (R\rho)_{[n/2]}$ and we fix such an inclusion for each ρ. If $R = \Pi\rho\, R\rho$ is the direct product of all the K-algebras $R\rho$, then R is a commutative K-algebra and we have a natural homomorphism $E \to R_{[n/2]}$ whose kernel is the intersection of the kernels of all the irreducible representations of E. Since E is semisimple, this intersection is trivial and the homomorphism is injective. The result follows.

It is clear from the above that if the algebra E is semisimple, then polynomial identities and bounded representation degrees are equivalent concepts.

Corollary 8.2 Let E_1 and E_2 be semisimple algebras over K. Suppose that E_i satisfies a nontrivial polynomial identity of degree n_i. Then $E_1 \otimes_K E_2$ satisfies the standard identity of degree $2[n_1/2][n_2/2]$.

Proof. By Theorem 8.1, $E_1 \subseteq (R^1)_{[n_1/2]}$ and $E_2 \subseteq (R^2)_{[n_2/2]}$ where R^1 and R^2 are commutative algebras over K. Then

$$\begin{aligned} E_1 \otimes_K E_2 &\subseteq R^1 \otimes_K K_{[n_1/2]} \otimes_K R^2 \otimes_K K_{[n_2/2]} \\ &= R^1 \otimes_K R^2 \otimes_K K_{[n_1/2][n_2/2]} \\ &= (R^1 \otimes_K R^2)_{[n_1/2][n_2/2]}. \end{aligned}$$

Thus again by Theorem 8.1, since $R^1 \otimes_K R^2$ is commutative, we see that $E_1 \otimes_K E_2$ satisfies the standard identity of degree $2[n_1/2][n_2/2]$.

It is not true that every algebra E which satisfies a polynomial identity or even a standard identity can be embedded in a full matrix ring over a commutative algebra. One such example is given below.

Theorem 8.3 (Amitsur [5]). Let K be a field. Then there exists a K-algebra E satisfying the standard identity $[\zeta_1, \zeta_2, \zeta_3, \zeta_4]$ of degree 4 which is not embeddable in a matrix ring over a commutative K-algebra of any degree.

Proof. Let $A = K[\xi, \eta]$ be the commutative polynomial ring over K in the two variables ξ and η subject only to the condition $\eta^3 = 0$. We consider A_2, the ring of 2×2 matrices over A. Define $\alpha, \beta \in A_2$ by

$$\alpha = \begin{pmatrix} 0 & \eta \\ \eta & 0 \end{pmatrix}, \qquad \beta = \begin{pmatrix} 1 & 0 \\ 0 & \xi \end{pmatrix}$$

and let B be the K-subalgebra of A_2 generated by 1, α, and β. Observe that for any integer $k \geq 0$, we have

$$\alpha\beta^k\alpha = \begin{pmatrix} \eta^2\xi^k & 0 \\ 0 & \eta^2 \end{pmatrix}.$$

For each integer $n \geq 1$, we define $I(n)$ to be the set of all matrices in B of the form

$$\begin{pmatrix} \eta^2 f(\xi) & 0 \\ 0 & \eta^2 g(\xi) \end{pmatrix}$$

with f and g polynomials in ξ and degree $f \leq n - 1$. Since $\eta^3 = 0$, it follows easily that $I(n)$ is an ideal in B. Moreover we have clearly $\alpha\beta^k\alpha \in I(n)$ for all $k < n$ but $\alpha\beta^n\alpha \notin I(n)$.

We claim now that $B/I(n)$ is not contained in the $n \times n$ matrix ring over any commutative K-algebra. Suppose by way of contradiction that $B/I(n) \subseteq C_n$ where C is commutative and let $\bar{\gamma}$ denote the image of $\gamma \in B$ in C_n. Since C is commutative, the matrix $\bar{\beta}$ satisfies its characteristic polynomial and hence by multiplying through by $\overline{\beta^0}$ if necessary we have

$$\overline{\beta^n} = \sum_{k<n} c_k \overline{\beta^k}$$

for suitable $c_k \in C$. Multiplying the above on the right and left by $\bar{\alpha}$ then yields

$$\overline{\alpha\beta^n\alpha} = \sum_{k<n} c_k \overline{\alpha\beta^k\alpha}.$$

Now for $k < n$, $\alpha\beta^k\alpha \in I(n)$, so $\overline{\alpha\beta^k\alpha} = 0$. Hence by the above $\overline{\alpha\beta^n\alpha} = 0$ so $\alpha\beta^n\alpha \in I(n)$, a contradiction.

Define E to be the direct product $E = \Pi_n\, B/I(n)$ of the K-algebras $B/I(n)$. Since E contains an isomorphic copy of each $B/I(n)$, it follows from the above that E is not embeddable in any matrix ring over a commutative K-algebra. Finally A_2 satisfies the standard identity s_4 and thus so does $B \subseteq A_2$, $B/I(n)$, and $E = \Pi_n\, B/I(n)$.

The following amusing result can be found in Ref. [*36*].

Theorem 8.4. Let $K[G]$ be a semisimple group ring which does not have r.b. n for some fixed integer n. Let G_n be the set of all elements $x \in G$ such that $\rho(x) = 1$ for all irreducible representations ρ of $K[G]$ of degree greater than n. Then G_n is a finite normal subgroup of G with $|G_n| \leq (2n)!$.

Proof. If $x, y \in G_n$ and ρ is an irreducible representation of degree greater than n, then $\rho(xy^{-1}) = \rho(x)\rho(y)^{-1} = 1$ and $xy^{-1} \in G_n$. If $x \in G_n$ and $y \in G$, then

$$\rho(y^{-1}xy) = \rho(y)^{-1}\rho(x)\rho(y) = \rho(y)^{-1}\rho(y) = 1$$

and $y^{-1}xy \in G_n$. Thus G_n is a normal subgroup of G.

By assumption, $K[G]$ is semisimple and does not have r.b. n. Thus by Theorem 8.1, $K[G]$ does not satisfy s_{2n}. Since s_{2n} is multilinear, it follows that there exists $g_1, g_2, \ldots, g_{2n} \in G$ with $\gamma = s_{2n}(g_1, g_2, \ldots, g_{2n}) \neq 0$. Let $x \in G_n$. We show that $\gamma(1 - x) = 0$. Let ρ be an irreducible representation of $K[G]$. If degree $\rho > n$, then $\rho(x) = 1 = \rho(1)$ so $\rho(\gamma(1 - x)) = 0$. If degree $\rho \leq n$, then by Lemma 4.3(iv), Theorem 4.6, and Theorem 6.4(iii), $\rho(K[G])$ satisfies s_{2n} so $\rho(\gamma) = 0$ and $\rho\big(\gamma(1 - x)\big) = 0$. Thus $\rho\big(\gamma(1 - x)\big) = 0$ for all ρ and $\gamma(1 - x) = 0$ since $K[G]$ is semisimple.

Thus $\gamma x = \gamma$ so $(\text{Supp } \gamma)x = \text{Supp } \gamma$. This says that G_n permutes the elements of $\text{Supp } \gamma \neq \varnothing$ by right multiplication and thus clearly $|G_n| \leq |\text{Supp } \gamma| \leq (2n)!$ since $\gamma = s_{2n}(g_1, g_2, \ldots, g_{2n})$ is a sum of $(2n)!$ monomials.

There are examples of groups G with $G_n \neq \langle 1 \rangle$. Let K be the field of complex numbers and let $G = Q_1 \times Q_2$ be the direct product of two copies of the quaternion group of order 8. Then $K[G]$ is semisimple and it has sixteen irreducible representations of degree 1, eight of degree 2, and precisely one of degree 4. Thus $K[G]$ does not have r.b. 2 and G_2 is easily seen to be the third subgroup of order 2 in $\mathbf{Z}(G) = \mathbf{Z}(Q_1) \times \mathbf{Z}(Q_2)$.

§9. LARGE CENTRALIZERS

We now begin our study of finite groups with r.b. n leading to the result in Section 12 that such groups have big abelian subgroups. The original proof of that result was entirely character theoretic in nature and applied only to fields of characteristic 0. A later characteristic p analogue depended on a trick to reduce the problem to the characteristic 0 situation. The main difficulty in finding a proof which is essentially independent of the field was that a replacement for Jordan's theorem on finite complex linear groups seemed to be needed. Such a replacement in characteristic 0 has recently appeared in Ref. [*45*] and we offer here a generalization of that argument which is applicable in all cases.

Let φ denote the natural projection $\varphi : K[G] \to K[Z]$ where $Z = \mathbf{Z}(G)$. Thus

$$\alpha = \sum_{x \in G} a_x x \to \varphi(\alpha) = \sum_{x \in Z} a_x x.$$

Lemma 9.1. Let G be a finite group and let $\alpha_1, \alpha_2, \ldots, \alpha_t, \beta_1, \beta_2, \ldots, \beta_t \in K[G]$ with

$$\Big(\bigcup_i \text{Supp } \alpha_i\Big) - Z = \{y_1, y_2, \ldots, y_r\},$$

$$\Big(\bigcup_i \text{Supp } \beta_i\Big) = \{z_1, z_2, \ldots, z_s\}.$$

Let T be a subset of G and suppose that

$$\alpha_1 x \beta_1 + \alpha_2 x \beta_2 + \cdots + \alpha_t x \beta_t = 0$$

for all $x \in T$. Then either $|\mathbf{C}_G(y_i)| \geq |T|/rs$ for some i or

$$\varphi(\alpha_1)\beta_1 + \varphi(\alpha_2)\beta_2 + \cdots + \varphi(\alpha_t)\beta_t = 0.$$

Proof. Suppose that

$$\gamma = \varphi(\alpha_1)\beta_1 + \varphi(\alpha_2)\beta_2 + \cdots + \varphi(\alpha_t)\beta_t \neq 0$$

and let $v \in \operatorname{Supp} \gamma$. Write $\alpha_i = \varphi(\alpha_i) + \alpha_i'$ and then write

$$\alpha_i' = \sum_{j=1}^{r} a_{ij} y_j, \qquad \beta_i = \sum_{j=1}^{s} b_{ij} z_j.$$

If y_i is conjugate to vz_j^{-1} in G for some i, j, choose $h_{ij} \in G$ with $h_{ij}^{-1} y_i h_{ij} = vz_j^{-1}$.

Let $x \in T$. Then since x centralizes $\varphi(\alpha_i)$ for all i, we have

$$\begin{aligned} 0 &= x^{-1}\alpha_1 x\beta_1 + x^{-1}\alpha_2 x\beta_2 + \cdots + x^{-1}\alpha_t x\beta_t \\ &= [\varphi(\alpha_1)\beta_1 + \varphi(\alpha_2)\beta_2 + \cdots + \varphi(\alpha_t)\beta_t] \\ &\quad + [(\alpha_1')^x\beta_1 + (\alpha_2')^x\beta_2 + \cdots + (\alpha_t')^x\beta_t]. \end{aligned}$$

Now v occurs in the support of the first term above which is γ and hence it must be canceled by something from the second term. Thus for some i, j we have $x^{-1}y_i x z_j = v$ or

$$x^{-1} y_i x = vz_j^{-1} = h_{ij}^{-1} y_i h_{ij}$$

and $x \in \mathbf{C}(y_i)h_{ij}$.

We have therefore shown that $T \subseteq \bigcup \mathbf{C}(y_i)h_{ij}$ and thus clearly

$$|T| \leq rs \cdot max_i\, |\mathbf{C}(y_i)|.$$

The result follows.

Lemma 9.2. Let G be finite and nonabelian and let $K[G]$ satisfy a polynomial identity of degree n. Then there exists an element $x \in G - \mathbf{Z}(G)$ with $[G:\mathbf{C}(x)] \leq (n!)^2$.

Proof. We assume by way of contradiction that $[G:\mathbf{C}(x)] > (n!)^2$ for all $x \in G - Z$ where $Z = \mathbf{Z}(G)$. By Lemma 4.1 we may assume that $K[G]$ satisfies the polynomial identity

$$f(\zeta_1, \zeta_2, \ldots, \zeta_n) = \zeta_1\zeta_2 \cdots \zeta_n + \sum_{\substack{\sigma \in S_n \\ \sigma \neq 1}} a_\sigma \zeta_{\sigma(1)}\zeta_{\sigma(2)} \cdots \zeta_{\sigma(n)}$$

so that clearly $n > 1$. For $j = 1, 2, \ldots, n$, define $f_j \in K[\zeta_j, \zeta_{j+1}, \ldots, \zeta_n]$ by

$$f = \zeta_1\zeta_2 \cdots \zeta_{j-1} f_j + \textit{terms not starting with } \zeta_1\zeta_2 \cdots \zeta_{j-1}.$$

Then clearly $f_1 = f$, $f_n = \zeta_n$, and f_j is a homogeneous multilinear polynomial of degree $n - j + 1$. In particular, for all j, ζ_j occurs in each monomial of f_j. We clearly have

$$f_j = \zeta_j f_{j+1} + \textit{terms not starting with } \zeta_j.$$

For $j = 2, 3, \ldots, n$, let $\mathcal{M}_j$ denote the set of all linear monomials in $K[\zeta_j, \zeta_{j+1}, \ldots, \zeta_n]$ and let $\mathcal{M}_1$ be empty. Then by Lemma 5.4 we have for all j, $|\mathcal{M}_j| \le |\mathcal{M}_2| \le n!$. We show now by induction on $j = 1, 2, \ldots, n$ that for any $x_j, x_{j+1}, \ldots, x_n \in G$, then either $f_j(x_j, x_{j+1}, \ldots, x_n) = 0$ or $\mu(x_j, x_{j+1}, \ldots, x_n) \in Z$ for some $\mu \in \mathcal{M}_j$. Since $f = f_1$ is a polynomial identity satisfied by $K[G]$, the result for $j = 1$ is clear.

Suppose the result holds for some $j < n$. Fix $x_{j+1}, x_{j+2}, \ldots, x_n \in G$ and let $x \subset G$ play the role of the jth variable. Let $\mu \in \mathcal{M}_{j+1}$. If $\mu(x_{j+1}, x_{j+2}, \ldots, x_n) \in Z$, we are done. Thus we may assume that $\mu(x_{j+1}, x_{j+2}, \ldots, x_n) \notin Z$ for all $\mu \in \mathcal{M}_{j+1}$. Set $\mathcal{M}_j - \mathcal{M}_{j+1} = \mathcal{T}_j$.

Now let $\mu \in \mathcal{T}_j$ so that μ involves the variable ζ_j. Write $\mu = \mu'\zeta_j\mu''$ where μ' and μ'' are monomials in $K[\zeta_{j+1}, \zeta_{j+2}, \ldots, \zeta_n]$. Then $\mu(x, x_{j+1}, \ldots, x_n) \in Z$ if and only if

$$x \in \mu'(x_{j+1}, \ldots, x_n)^{-1} Z \mu''(x_{j+1}, \ldots, x_n)^{-1} = Zh_\mu,$$

a fixed coset of Z, since μ' and μ'' do not involve ζ_j. Thus it follows that for all $x \in T = G - \bigcup_{\mu \in \mathcal{T}_j} Zh_\mu$, we have $\mu(x, x_{j+1}, \ldots, x_n) \notin Z$ for all $\mu \in \mathcal{M}_j$ since $\mathcal{M}_j \subseteq \mathcal{M}_{j+1} \cup \mathcal{T}_j$. Since the inductive result holds for j, we conclude that for all $x \in T$ we have $f_j(x, x_{j+1}, \ldots, x_n) = 0$.

Write

$$f_j(\zeta_j, \zeta_{j+1}, \ldots, \zeta_n) = \zeta_j f_{j+1} + \sum_t \eta_t \zeta_j \eta'_t$$

where $\eta_t, \eta'_t \in K[\zeta_{j+1}, \zeta_{j+2}, \ldots, \zeta_n]$ and η_t is a linear monomial. Hence $\eta_t \in \mathcal{M}_{j+1}$. Now by the above we have

$$0 = 1 \cdot x \cdot f_{j+1}(x_{j+1}, \ldots, x_n) + \sum_t \eta_t(x_{j+1}, \ldots, x_n) \cdot x \cdot \eta'_t(x_{j+1}, \ldots, x_n)$$

for all $x \in T$. We apply Lemma 9.1 and there are two possible conclusions.

Now f has at most $n!$ monomials and thus clearly in the notation of Lemma 9.1 we have $r \le n! - 1$, $s \le n!$, and $|T| \ge |G| - |\mathcal{T}_j|\,|Z| \ge |G| - (n!)\,|Z|$. Suppose $|\mathbf{C}(y_i)| \ge |T|/rs$ for some $y_i \in G - Z$. Then

$$(n!)^2\,|\mathbf{C}(y_i)| \ge |G| - (n!)\,|Z| + (n!)\,|\mathbf{C}(y_i)| \ge |G|$$

since $\mathbf{C}(y_i) \supseteq Z$. Thus $(n!)^2 \ge [G : \mathbf{C}(y_i)]$, a contradiction by assumption. Hence Lemma 9.1 implies that

$$0 = \varphi(1) f_{j+1}(x_{j+1}, \ldots, x_n) + \sum_t \varphi\big(\eta_t(x_{j+1}, \ldots, x_n)\big)\eta'_t(x_{j+1}, \ldots, x_n).$$

Clearly $\varphi(1) = 1$ and $\eta_t(x_{j+1}, \ldots, x_n) \in G - Z$ since $\eta_t \in \mathscr{M}_{j+1}$. Thus $\varphi(\eta_t(x_{j+1}, \ldots, x_n)) = 0$ and

$$0 = 1 \cdot f_{j+1}(x_{j+1}, \ldots, x_n) = f_{j+1}(x_{j+1}, \ldots, x_n)$$

and the induction step is proved.

In particular the inductive result holds for $j = n$. Here $f_n(\zeta_n) = \zeta_n$ and $\mathscr{M}_n = \{\zeta_n\}$. Thus we conclude that for all $x \in G$ we have either $x = 0$ or $x \in Z$, a contradiction since $G \neq Z$. Therefore the assumption that there exists no $x \in G - Z$ with $(n!)^2 \geq [G:\mathbf{C}(x)]$ is false and the lemma is proved.

We now obtain two useful consequences of the above.

Lemma 9.3. Let $G \neq \langle 1 \rangle$ be a finite group and let $K[G]$ satisfy a polynomial identity of degree n. Then G has a normal subgroup N such that $\mathbf{C}_G(N) \neq \langle 1 \rangle$ and

$$[G:N] \leq (n!)^{2(n!)^2}.$$

Proof. Since $G \neq \langle 1 \rangle$ it follows from Lemma 9.2 that there exists an element $x \in G$, $x \neq 1$ with $[G:\mathbf{C}(x)] \leq (n!)^2$. Let N be the intersection of the at most $(n!)^2$ conjugates of $\mathbf{C}(x)$. Then N is normal in G and $[G:N] \leq (n!)^{2(n!)^2}$ by Lemma 1.1. Since $x \in \mathbf{C}(N)$ and $x \neq 1$, the result follows.

Lemma 9.4. Let G be a finite group and let $K[G]$ satisfy a polynomial identity of degree n. Let q be a prime with $q > (n!)^2$. Then G has a normal abelian Sylow q-subgroup.

Proof. By induction on $|G|$. If G is abelian and in particular if $|G| = 1$, then the result is certainly true. Thus we may assume that G is nonabelian.

By Lemma 9.2 there exists an element $x \in G - \mathbf{Z}(G)$ such that if $H = \mathbf{C}(x)$, then $1 < [G:H] \leq (n!)^2 < q$. Let Q be a Sylow q-subgroup of G. Then from $|G| \geq |QH| = |H|\,[Q:Q \cap H]$, we have $q > [G:H] \geq [Q:Q \cap H]$ and hence $[Q:Q \cap H] = 1$ and $Q \subseteq H$. Thus all Sylow q-subgroups of G are contained in H. By Lemma 6.6, $K[H]$ satisfies a polynomial identity of degree n and hence by induction we conclude that H has a unique Sylow q-subgroup which is also abelian. Thus G has a unique Sylow q-subgroup which is therefore normal.

§10. MODULES

Let K be an algebraically closed field and let G be a finite group. If V is an irreducible $K[G]$-module, then the map $\rho: K[G] \to \mathrm{End}(V)$ is a homomorphism of $K[G]$ onto K_d, the full ring of $d \times d$ matrices over K. Here $d = \dim_K V = \dim V$ and thus $K[G]$ has r.b. n if and only if every irreducible $K[G]$-module has K-dimension at most n. In the following, all groups are assumed to be finite, all modules are assumed to be finite dimensional, and K is algebraically closed.

Let V be a $K[G]$-module and fix a basis for V. With respect to this basis, each element of $K[G]$ can of course be represented as a matrix and for each $\alpha \in K[G]$ we let $\chi_V(\alpha)$ denote the trace of its corresponding matrix. If a different basis for V is chosen, then the two different matrices for α are similar and hence have the same trace. Thus χ_V is a well-defined K-valued function on $K[G]$ which is called the character of V.

If V and W are two $K[G]$-modules, then we say that V and W are equivalent and write $V \simeq_G W$ or $V \simeq W$ if there exists a $K[G]$-isomorphism $\mu: V \to W$. Suppose that $V \simeq W$ with the isomorphism $\mu: V \to W$ and let $\{v_1, v_2, \ldots, v_d\}$ be a basis for V. Then $\{\mu(v_1), \mu(v_2), \ldots, \mu(v_d)\}$ is a basis for W and with respect to these corresponding bases the matrix representations of $K[G]$ on V and W are identical. Thus $\chi_V = \chi_W$.

Let U, V, and W be three $K[G]$-modules. Then $V \otimes_K W = V \otimes W$ becomes a $K[G]$-module under the diagonal action

$$(v \otimes w)x = (vx) \otimes (wx)$$

for $v \in V$, $w \in W$, $x \in G$. We have clearly

$$V \otimes W \simeq W \otimes V$$

$$U \otimes (V \otimes W) \simeq (U \otimes V) \otimes W$$

$$V \otimes W \simeq V' \otimes W' \quad \textit{if} \quad V \simeq V', \quad W \simeq W'.$$

A module V is said to be linear if $\dim V = 1$. The corresponding representation is then clearly given by $\alpha \to [\chi_V(\alpha)]$. Thus χ_V induces a multiplicative homomorphism of G into $K - \{0\}$ and the equivalence class of V is uniquely determined by χ_V. Certainly V is irreducible. We let V_0 denote the principal $K[G]$-module so that V_0 is linear and $\chi_{V_0}(x) = 1$ for all $x \in G$. If $\dim V > 1$, then V is said to be nonlinear.

If V and W are $K[G]$-modules, we use $V \leftrightarrow W$ to indicate that V and W have equivalent composition factors. If $V_1, V_2, \ldots, V_t$ are the composition factors of V, then clearly $V \leftrightarrow V_1 + V_2 + \cdots + V_t$ and the V_i are uniquely determined up to equivalence by the Jordan–Holder theorem.

Lemma 10.1.

(i) If $|G| \neq 0$ in K, then $K[G]$ is a finite dimensional semisimple algebra and every $K[G]$-module is a direct sum of irreducible $K[G]$-modules. If, in addition, G is abelian, then $K[G]$ has precisely $|G|$ inequivalent linear modules.

(ii) Let P be a p-group and let K have characteristic p. Then

$$JK[P] = \{\sum a_x x \in K[P] \mid \sum a_x = 0\}.$$

(iii) $K[G]$ has r.b. 1 if and only if either G is abelian or K has characteristic p and G' is a p-group.

Proof. (i) By Theorems 3.3 and 3.6, $K[G]$ has no nonzero nilpotent ideals. The first result therefore follows from the Wedderburn theorems since $K[G]$ is a finite dimensional algebra. Now $K[G]$ is a ring direct sum of full matrix rings over K, each corresponding to an equivalence class of irreducible $K[G]$-modules. If G is abelian, then $K[G]$ is commutative and so these matrix rings are all of degree 1 and there are precisely $|G| = \dim_K K[G]$ of these.

(ii) Set $R = \{\sum a_x x \in K[G] \mid \sum a_x = 0\}$. Since R is clearly a maximal ideal in $K[P]$, it suffices to show that R is a radical ring. If the latter is not the case, then R has an irreducible representation and thus there is a homomorphism ρ of R onto a full matrix ring K_m. Now R is spanned over K by the elements $1 - x$ with $x \in P$ and each of these is easily seen to be nilpotent since if $x^{p^r} = 1$, then $(1 - x)^{p^r} = 1 - x^{p^r} = 0$. Thus the matrices $\rho(1 - x)$ are nilpotent and hence have trace zero. This therefore implies that every element of $\rho(R) = K_m$ has trace zero, a contradiction.

(iii) If K has characteristic 0 then by (i) $K[G]$ has r.b. 1 if and only if $K[G]$ is commutative and hence if and only if G is abelian. Now let K have characteristic $p > 0$. Suppose first that $K[G]$ has r.b. 1. Then $K[G]/JK[G]$ is a direct sum of 1×1 matrix rings over K and hence is commutative. Thus for $x, y \in G$ we have $xy - yx \in JK[G]$ and hence since $JK[G]$ is an ideal

$$1 - y^{-1}x^{-1}yx = y^{-1}x^{-1}(xy - yx) \in JK[G].$$

Let $H = \{h \in G \mid 1 - h \in JK[G]\}$. It follows easily that H is a subgroup of G and since $y^{-1}x^{-1}yx \in H$ for all $x, y \in G$ we see that $H \supseteq G'$. Let $h \in H$. Then $1 - h \in JK[G]$ and $JK[G]$ is nilpotent, so $(1 - h)^s = 0$ for some positive integer s. If $p^r \geq s$, then $0 = (1 - h)^{p^r} = 1 - h^{p^r}$ and $h^{p^r} = 1$. Thus H is a p-group and hence so is G'.

Finally suppose that $P = G'$ is a p-group. Since P is normal in G, it follows easily from (ii) that $K[G] \cdot JK[P] = JK[P] \cdot K[G]$ and thus

$K[G] \cdot JK[P]$ is a nilpotent ideal in $K[G]$. Therefore $K[G] \cdot JK[P] \subseteq JK[G]$. Now by (ii) the kernel of the natural epimorphism $K[G] \to K[G/G']$ is clearly $K[G] \cdot JK[P]$ and hence every irreducible $K[G]$-module is in fact an irreducible $K[G/G']$-module. Since G/G' is abelian, $K[G/G']$ has r.b. 1 and hence so does $K[G]$.

Lemma 10.2. Let H be a subgroup of G.

(i) If W is an irreducible $K[H]$-module, then there exists an irreducible $K[G]$-module V such that W is a submodule of V_H, the restriction of V to $K[H]$.

(ii) If $K[G]$ has r.b. n, then so does $K[H]$ and also $K[G/H]$ if H is normal in G.

Proof. (i) Since W is irreducible, there exists a maximal right ideal M of $K[H]$ such that $K[H]/M \simeq {}_H W$. Consider the right ideal $M \cdot K[G]$ of $K[G]$. If $M \cdot K[G] = K[G]$, then there exists $\alpha_i \in M$, $\beta_i \in K[G]$ with $\sum \alpha_i \beta_i = 1$. For each i write $\beta_i = \beta_i' + \beta_i''$ where Supp $\beta_i' \subseteq H$ and (Supp β_i'') $\cap H = \varnothing$. Then

$$(\textstyle\sum \alpha_i \beta_i') + (\sum \alpha_i \beta_i'') = 1.$$

Since Supp $\sum \alpha_i \beta_i' \subseteq H$ and since (Supp $\sum \alpha_i \beta_i''$) $\cap H = \varnothing$, we conclude that $1 = \sum \alpha_i \beta_i' \in M$, a contradiction. Thus $M \cdot K[G]$ is a proper right ideal of $K[G]$; and let N be a maximal right ideal with $N \supseteq M \cdot K[G]$. Since $1 \notin N$, we have $K[H] > N \cap K[H] \supseteq M$ so $N \cap K[H] = M$. Let V be the irreducible $K[G]$-module $K[G]/N$. Then V_H has as a submodule

$$(N + K[H])/N \simeq_H K[H]/(N \cap K[H]) = K[H]/M \simeq_H W$$

and this result follows.

(ii) By (i), if W is any irreducible $K[H]$-module, then dim $W \leq$ dim V for some irreducible $K[G]$-module V. Since dim $V \leq n$, $K[H]$ has r.b. n. If H is normal in G, then every irreducible $K[G/H]$-module becomes an irreducible $K[G]$-module by way of the epimorphism $K[G] \to K[G/H]$. Thus $K[G/H]$ has r.b. n.

Lemma 10.3. Let V, W, and L be $K[G]$-modules with L linear.

(i) V is irreducible if and only if $V \otimes L$ is irreducible.

(ii) $\chi_{V \otimes L}(x) = \chi_V(x)\chi_L(x)$ for all $x \in G$.

(iii) If $V \leftrightarrow W$, then $V \otimes L \leftrightarrow W \otimes L$.

Proof. (i) If W is a submodule of V, then $W \otimes L = W'$ is a submodule of $V \otimes L$. Conversely suppose that W' is a submodule of $V \otimes L$.

Let u be a fixed nonzero element of L. Then every element of $V \otimes L$ can be written uniquely as $v \otimes u$ for some $v \in V$. It then follows easily that $W = \{v \in V \mid v \otimes u \in W'\}$ is a submodule of V.

(ii) Let u be a fixed element of L, $u \neq 0$, and let $\{v_1, v_2, \ldots, v_d\}$ be a basis for V. Then $\{v_1 \otimes u, v_2 \otimes u, \ldots, v_d \otimes u\}$ is a basis for $V \otimes L$. Let $x \in G$ and suppose that $v_i x = \sum a_{ij} v_j$ with $a_{ij} \in K$. Then

$$(v_i \otimes u)x = (v_i x) \otimes (ux) = \left(\sum_j a_{ij} v_j\right) \otimes (\chi_L(x)u)$$
$$= \sum_j a_{ij}\chi_L(x) \cdot v_j \otimes u.$$

Therefore

$$\chi_{V\otimes L}(x) = \sum_i a_{ii}\chi_L(x) = \chi_V(x)\chi_L(x).$$

(iii) Let $V_1, V_2, \ldots, V_t$ be the composition factors of V. It clearly suffices to show that $V_1 \otimes L, V_2 \otimes L, \ldots, V_t \otimes L$ are the composition factors of $V \otimes L$. By (i) these are at least irreducible. Say V_1 is a submodule of V. Then $V_1 \otimes L$ is a submodule of $V \otimes L$. Let u be a fixed nonzero element of L. Then the map $\eta: V \otimes L \to (V/V_1) \otimes L$ defined by $\eta(v \otimes u) = (v + V_1) \otimes u$ is easily seen to be a $K[G]$-epimorphism with kernel $V_1 \otimes L$. Thus $(V \otimes L)/(V_1 \otimes L) \simeq_G (V/V_1) \otimes L$. The result follows by induction on the number t of composition factors.

Lemma 10.4. Let V be an irreducible $K[G]$-module and let H be a normal subgroup of G. Suppose that $\chi_V(x) = 0$ for all $x \in G - H$. Then

(i) $[G:H] \leq (\dim V)^2$.

(ii) If $H \neq G$, then V_H is reducible.

Proof. Let ρ denote the corresponding epimorphism $\rho: K[G] \to K_d \simeq \operatorname{End}(V)$. Since some element of $\rho(K[G])$ has nonzero trace, it follows that for some $h \in G$ we have $\chi_V(h) \neq 0$. By assumption we must have $h \in H$.

Let $x_1, x_2, \ldots, x_t$ be a complete set of coset representatives for H in G and suppose that $\sum_i a_i\rho(x_i) = 0$ with $a_i \in K$. Fix j and multiply this expression on the right by $\rho(x_j^{-1}h)$ to obtain $\sum_i a_i\rho(x_i x_j^{-1}h) = 0$. Now for $i \neq j$, $x_i x_j^{-1}h \notin H$ so by assumption $0 = \chi_V(x_i x_j^{-1}h) = \operatorname{trace} \rho(x_i x_j^{-1}h)$. Thus taking traces of the above yields $a_j\chi_V(h) = a_j\chi_V(x_j x_j^{-1}h) = 0$ and hence $a_j = 0$ since $\chi_V(h) \neq 0$. We have therefore shown that $\rho(x_1)$, $\rho(x_2), \ldots, \rho(x_t)$ are linearly independent. Since $\dim_K \operatorname{End}(V) = d^2$ where $d = \dim_K V$, we conclude that $[G:H] = t \leq d^2 = (\dim V)^2$ and (i) follows.

Suppose now that V_H is irreducible. Then $\rho(K[H]) = K_d$. If $H \neq G$, we can choose $x \in G - H$. Then clearly $\rho(K[H]x) = K_d$. Now $K[H]x$ is spanned by all elements $y \in Hx$ and for each such y we have trace $\rho(y) = \chi_V(y) = 0$ by assumption. This yields trace $K_d =$ trace $\rho(K[H]x) = 0$, certainly a contradiction. Thus V_H is reducible.

The following lemma is crucial to the proof of our main result on finite groups.

Lemma 10.5. Let $K[G]$ have r.b. n and let H be normal in G. Let W be an irreducible $K[H]$-module with dim $W > n/2$. Then

(i) If V is an irreducible $K[G]$-module such that W is a submodule of V_H, then $W = V_H$.

(ii) There exists an irreducible $K[G]$-module V with $W = V_H$.

(iii) $K[G/H]$ has r.b. 1.

Proof. (i) Suppose $W \subseteq V$ and that W is invariant under $K[H]$. Let $x \in G$ and consider $Wx \subseteq V$. Since H is normal in G, we have

$$(Wx)K[H] = W(xK[H]) = W(K[H]x) = Wx$$

so that Wx is also a $K[H]$-submodule of V_H. Since $W = (Wx)x^{-1}$ it follows that Wx is also irreducible and hence either $W = Wx$ or $W \cap Wx = 0$. If $W \cap Wx = 0$, then $W + Wx$ is a direct sum and

$$n \geq \dim V \geq \dim W + \dim Wx = 2 \dim W > n,$$

a contradiction. Thus $W = Wx$ and since this is true for all $x \in G$ we see that W is a $K[G]$-submodule of V. Now V is irreducible so $W = V$.

(ii) By Lemma 10.2(i) there exists an irreducible $K[G]$-module V such that W is a submodule of V_H. By (i) above $V_H = W$.

(iii) Let V be as in (ii) and let U be an irreducible $K[G/H]$-module which we view as an irreducible $K[G]$-module. Consider the $K[G]$-module $V \otimes U$. Let B be a nonzero submodule of $V \otimes U$ and let $\bar{v} = \sum_1^d v_i \otimes u_i$ be a nonzero element of B. Here $\{v_1, v_2, \ldots, v_d\}$ is a basis for V and $\{u_i\} \subseteq U$. Since $\bar{v} \neq 0$, say $u_1 \neq 0$. Let $v \in V$ be arbitrary. Since $V_H = W$ is irreducible, there exists $\alpha = \sum a_x x \in K[H]$ with $v_1\alpha = v$, $v_j\alpha = 0$ for $j \neq 1$. Now for $x \in H$ we have $u_i x = u_i$ so

$$\begin{aligned}(v_i \otimes u_i)\alpha &= \sum_x a_x (v_i \otimes u_i)x \\ &= \sum_x (a_x v_i x) \otimes u_i = v_i\alpha \otimes u_i\end{aligned}$$

and thus $v \otimes u_1 = \bar{v}\alpha \in B$. We have therefore shown that $V \otimes u_1 \subseteq B$ and hence if $U' = \{u \in U \mid V \otimes u \subseteq B\}$, then $U' \neq 0$. It follows easily that U' is a submodule of U and since U is irreducible we have $U' = U$ and $B = V \otimes U$. Thus $V \otimes U$ is irreducible so

$$n \geq \dim (V \otimes U) = (\dim V)(\dim U) > (n/2) \dim U$$

and $\dim U = 1$. Therefore $K[G/H]$ has r.b. 1.

§11. REDUCTION IN SPECIAL CASES

We continue with the assumptions of the preceding section. Thus K is an algebraically closed field, all groups are finite, and all modules are finite dimensional.

A group E is said to be of type (*) if

(1) E is nonabelian and $|E'| \neq 0$ in K.

(2) All proper homomorphic images of E are abelian.

(3) E is nilpotent of class 2.

Lemma 11.1. Let E be a group of type (*) and let U be a nonlinear irreducible $K[E]$-module. Then

(i) $[E:\mathbf{Z}(E)] = (\dim U)^2$.

(ii) $K[E/\mathbf{Z}(E)]$ has $|E/\mathbf{Z}(E)|$ inequivalent linear modules and for any such L we have $U \otimes L \simeq_E U$.

Proof. Let Y be a central subgroup of E of order q for some prime q. Then by (2), E/Y is abelian so we must have $E' = Y$ cyclic of prime order q and K does not have characteristic q. Let $Z = \mathbf{Z}(E)$.

Let ρ denote the homomorphism $\rho: K[E] \to K_d \simeq \mathrm{End}(U)$ where $d = \dim U$. If $z \in Z$, then $\rho(z)$ must be a scalar matrix, so say $\rho(z) = \lambda(z)\rho(1)$ where $\lambda(z) \in K$ and $\rho(1)$ is of course the identity matrix. It follows easily that $\lambda: Z \to K - \{0\}$ is a multiplicative homomorphism. Let $z \in E'$, $z \neq 1$, so that $E' = \langle z \rangle$. If $\lambda(z) = 1$, then every element of E' is mapped to the identity under ρ and this implies easily that U is in fact an irreducible $K[E/E']$-module, a contradiction by Lemma 10.1 since U is nonlinear. Thus $\lambda(z) \neq 1$ for all $z \in E'$, $z \neq 1$.

Now let $x \in E - Z$. Since x is not central, there exists an element $y \in E$ with $z = x^{-1}y^{-1}xy \neq 1$. Then $z \in E' \subseteq Z$ and $y^{-1}xy = xz$ so we have

$$\rho(y)^{-1}\rho(x)\rho(y) = \rho(x)\rho(z) = \lambda(z)\rho(x).$$

Since similar matrices have the same trace, this yields

$$\begin{aligned}\chi_U(x) &= \mathrm{trace}\ \rho(x) = \mathrm{trace}\ \rho(y)^{-1}\rho(x)\rho(y) \\ &= \mathrm{trace}\ \lambda(z)\rho(x) = \lambda(z)\chi_U(x)\end{aligned}$$

and hence $\chi_U(x) = 0$ since $\lambda(z) \neq 1$. Thus by Lemma 10.4, $[E:Z] \leq d^2$.

Now the group homomorphism $\lambda: Z \to K - \{0\}$ induces an algebra epimorphism $K[Z] \to K$; let I_0 be the kernel of this map. Then I_0 is an ideal in $K[Z]$ of codimension 1. Set $I = I_0 \cdot K[E]$ so that I is an ideal since I_0 is central. If $x_1, x_2, \ldots, x_t$ are a complete set of coset representatives for Z in E where $t = [E:Z]$, then clearly $I = I_0x_1 + I_0x_2 + \cdots + I_0x_t$ is a vector space direct sum. Thus $K[E]/I$ is an algebra of dimension t. Clearly $\rho(I) = 0$ and hence ρ induces an epimorphism $\rho': K[E]/I \to K_d$. However by the above

$$\dim K[E]/I = t \leq d^2 = \dim K_d$$

so ρ' must in fact be an isomorphism and $t = d^2$. This yields (i).

We remark that U exists since $K[E]$ does not have r.b. 1. by Lemma 10.1(iii) and assumption (1) above. Now some element of $\rho(K[E]) = K_d$ has nonzero trace and hence $\chi_U(x) \neq 0$ for some $x \in E$. By the above we must have $x \in Z$ and then $\chi_U(x) = d\lambda(x) \neq 0$. Thus $d \neq 0$ in K and $|E/\mathbf{Z}(E)| = d^2 \neq 0$ in K. By Lemma 10.1(i), $K[E/\mathbf{Z}(E)]$ has $|E/\mathbf{Z}(E)|$ inequivalent linear modules.

Let L be a linear module of $K[E/Z]$. Then by Lemma 10.3, $U' = U \otimes L$ is also a nonlinear irreducible $K[E]$-module. If $u \in U$ and $v \in L$, then for all $z \in Z$ we have $(u \otimes v)z = (uz) \otimes (vz) = \lambda(z)(u \otimes v)$ since $vz = v$. It then follows that U' is also an irreducible module of $K[E]/I \simeq K_d$. Since all irreducible modules of K_d are isomorphic, we have $U \simeq_E U'$ and the result follows.

Lemma 11.2. Let $K[G]$ have r.b. n and let H be normal in G with $G/H = E$, a group of type (*). Let $G \supseteq Z \supseteq H$ with $Z/H = \mathbf{Z}(E)$. Then $[G:Z] \leq n^2$ and $K[Z]$ has r.b. $[n/2]$.

Proof. Let U be a nonlinear irreducible $K[E]$-module viewed as one of $K[G]$. Such a module exists by Lemma 10.1(iii). Then $\dim U \leq n$ and so by Lemma 11.1(i) we have

$$[G:Z] = [E:\mathbf{Z}(E)] = (\dim U)^2 \leq n^2.$$

Clearly $U_Z = M_1 + M_2 + \cdots + M_d$, the direct sum of $d = \dim U$ equivalent linear modules.

Suppose by way of contradiction that $K[Z]$ does not have r.b. $[n/2]$ and let W be an irreducible $K[Z]$-module with $\dim W > n/2$. By Lemma 10.5(ii) there exists an irreducible $K[G]$-module V with $V_Z = W$. We consider $V \otimes U$. Say $V \otimes U \leftrightarrow V_1 + V_2 + \cdots + V_t$, a direct sum of t irreducible $K[G]$-modules. Then clearly

$$\begin{aligned}(V_1)_Z + (V_2)_Z + \cdots + (V_t)_Z \\ &\leftrightarrow (V \otimes U)_Z \\ &= W \otimes (M_1 + M_2 + \cdots + M_d) \\ &= (W \otimes M_1) + (W \otimes M_2) + \cdots + (W \otimes M_d)\end{aligned}$$

Note that for each j, $W \otimes M_j$ is an irreducible $K[Z]$-module with $\dim (W \otimes M_j) = \dim W > n/2$. Now for each i, $(V_i)_Z$ must have an isomorphic copy of some $W \otimes M_j$ as a submodule by the Jordan–Holder theorem and hence by Lemma 10.5(i), $(V_i)_Z \simeq_Z W \otimes M_j$ for some j. It then follows that $t = d$.

Let L be a linear module of $K[E/\mathbf{Z}(E)]$ viewed as one of $K[G]$. Then by Lemma 10.3(iii)

$$\begin{aligned}(V_1 \otimes L) + (V_2 \otimes L) + \cdots + (V_d \otimes L) &\leftrightarrow (V \otimes U) \otimes L \\ &= V \otimes (U \otimes L) \simeq V \otimes U \\ &\leftrightarrow V_1 + V_2 + \cdots + V_d\end{aligned}$$

by Lemma 11.1(ii). Since $V_1 \otimes L$ is irreducible, it follows from the Jordan–Holder theorem that $V_1 \otimes L \simeq_G V_{i(L)}$ where $i(L)$ is some subscript depending on L. Now there are only d possible subscripts but there are $d^2 = |E/\mathbf{Z}(E)|$ possible L's by Lemma 11.1. Since $d^2 > d$, there exist two inequivalent L's say L_1 and L_2 such that $V_1 \otimes L_1 \simeq_G V_1 \otimes L_2$. By Lemma 10.3(ii), we have for all $x \in G$

$$\begin{aligned}\chi_{V_1}(x)\chi_{L_1}(x) &= \chi_{V_1 \otimes L_1}(x) \\ &= \chi_{V_1 \otimes L_2}(x) = \chi_{V_1}(x)\chi_{L_2}(x).\end{aligned}$$

Let $\bar{H} = \{x \in G \mid \chi_{L_1}(x) = \chi_{L_2}(x)\}$. Since χ_{L_1} and χ_{L_2} are homomorphisms, we see that $\bar{H}$ is a subgroup of G and also $\bar{H} \supseteq Z$ since for $x \in Z$, $\chi_{L_1}(x) = 1 = \chi_{L_2}(x)$. On the other hand, L_1 and L_2 are inequivalent so $\bar{H} \neq G$. Now if $x \notin \bar{H}$, then $\chi_{V_1}(x)\chi_{L_1}(x) = \chi_{V_1}(x)\chi_{L_2}(x)$ implies that $\chi_{V_1}(x) = 0$. Hence by Lemma 10.4(ii), we conclude that $(V_1)_{\bar{H}}$ is reducible. Since $\bar{H} \supseteq Z$ and $(V_1)_Z \simeq W \otimes M_j$ is irreducible, we have a contradiction and the result follows.

Lemma 11.3. Suppose E is a group, A is a normal abelian subgroup of E, and $E = \langle A, x\rangle$ for some $x \in E$. Suppose further that A is self-centralizing, that is, $\mathbf{C}_E(A) = A$. Then

(i) $E' = (x, A) = \{(x, a) = x^{-1}a^{-1}xa \mid a \in A\}$.

(ii) $\mathbf{C}_A(x) = \mathbf{Z}(E)$.

(iii) $|E'| \cdot |\mathbf{Z}(E)| = |A|$.

Proof. Since A is abelian, the map $a \to (x^{-1}a^{-1}x)a$ is the product of two endomorphisms of A and hence is an endomorphism of A. The image of the map is (x, A) which is therefore a subgroup and the kernel is clearly $\mathbf{C}_A(x)$. Thus $|(x, A)| \cdot |\mathbf{C}_A(x)| = |A|$.

Since A is self-centralizing, we have $A \supseteq \mathbf{Z}(E)$ and hence $\mathbf{C}_A(x) = A \cap \mathbf{C}_E(x) \supseteq \mathbf{Z}(E)$. On the other hand, $\mathbf{C}_A(x)$ centralizes both A and x and hence $\mathbf{C}_A(x) = \mathbf{Z}(E)$ since $E = \langle A, x\rangle$.

The result will follow if we show that $E' = (x, A)$. Clearly $E' \supseteq (x, A)$. Now $A \supseteq (x, A)$ and A is abelian so A normalizes (x, A). Also $(x, A)^x = (x, A^x) = (x, A)$ so x normalizes (x, A) and hence (x, A) is normal in $\langle x, A\rangle = E$. Now $\bar{E} = E/(x, A)$ is generated by the abelian group $\bar{A} = A/(x, A)$ and the element $\bar{x} = x(x, A)$. Furthermore $\bar{x}$ centralizes $\bar{A}$ so $\bar{E}$ is abelian and $(x, A) \supseteq E'$. Thus $(x, A) = E'$ and the lemma is proved.

A group E is said to be of type (**) if

(1) E is nonabelian and $|E'| \neq 0$ in K.

(2) All proper homomorphic images of E are abelian.

(3) E has a nonidentity normal abelian subgroup.

(4) E is not nilpotent of class 2.

Lemma 11.4. Let E be a group of type (**) and let U be a nonlinear irreducible $K[E]$-module. Then E' is a normal abelian subgroup of E and

(i) $[E:E'] \leq \dim U$.

(ii) $\mathbf{C}_{E'}(x) = \langle 1\rangle$ for all $x \in E - E'$.

Proof. Let B be the nonidentity normal abelian subgroup of E given by (3). Then E/B is abelian by (2) and hence $B \supseteq E'$. Also every subgroup H of E with $E \supseteq H \supseteq B$ is normal. Choose $A \supseteq B$ such that A is abelian and maximal with this property. Clearly A is self-centralizing.

Let $x \in E - A$ and set $H = \langle A, x\rangle$. Now H is nonabelian by the choice of A and since H is normal in E, we conclude that H' is a nonidentity normal subgroup of E and thus $H' \supseteq E'$ by (2). Therefore $E' = H' = (x, A)$ by Lemma 11.3. Also by that lemma, we have $|H'| \cdot |\mathbf{Z}(H)| = |A|$ so $|\mathbf{Z}(H)|$ is the same for all choices of $x \in E - A$.

Since E is not nilpotent of class 2, it follows that there exists $y \in E$ such that y does not centralize E'. Clearly $y \in E - A$; set $\bar{H} = \langle A, y \rangle$. If $\mathbf{Z}(\bar{H}) \neq \langle 1 \rangle$, then since $\mathbf{Z}(\bar{H})$ is normal in E, we have $\mathbf{Z}(\bar{H}) \supseteq E'$ and hence y centralizes E', a contradiction. Thus $\mathbf{Z}(\bar{H}) = \langle 1 \rangle$ and by the above $|\mathbf{Z}(H)| = 1$ for all $x \in E - A$. Since $|E'| \cdot |\mathbf{Z}(H)| = |A|$ and $A \supseteq E'$, we conclude that $A = E' = (A, x)$ for all $x \in E - A$. Furthermore by Lemma 11.3, $\mathbf{C}_A(x) = \mathbf{Z}(H) = \langle 1 \rangle$ and (ii) follows.

Now by Lemma 10.1 and the fact that $E' = A$ is abelian and $|E'| \neq 0$ in K, we have $U_A = L_1 + L_2 + \cdots + L_d$, a direct sum of linear $K[A]$-modules. If each L_i is principal, then each element of A acts like the identity and U is in fact an irreducible $K[E/E']$-module, a contradiction since U is nonlinear. Thus say $L = L_1$ is nonprincipal. Let $x_1, x_2, \ldots, x_t$ be a complete set of coset representatives for A in E. Since A is normal, we have $Lx_i \cdot K[A] = L \cdot K[A]x_i = Lx_i$ so Lx_i is also a $K[A]$-submodule of U_A. If $u \in L - \{0\}$ and $a \in A$, then

$$(ux_i)a = u(x_i a x_i^{-1})x_i = \chi_L(x_i a x_i^{-1})ux_i$$

and hence $\chi_{Lx_i}(a) = \chi_L(x_i a x_i^{-1})$.

Suppose $Lx_i \simeq_A Lx_j$. Then $\chi_L(x_i a x_i^{-1}) = \chi_L(x_j a x_j^{-1})$ for all $a \in A$ and since $\chi_L : A \to K - \{0\}$ is a multiplicative homomorphism, setting $b = x_i a x_i^{-1}$, $y = x_i x_j^{-1}$ gives $\chi_L(y^{-1}b^{-1}yb) = 1$ for all $b \in A$. If $i \neq j$, then $y \in E - A$ and hence $(y, A) = A$, $\chi_L(A) = 1$, and L is the principal $K[A]$-module, a contradiction. Thus the t modules $Lx_1, Lx_2, \ldots, Lx_t$ are inequivalent and by the Jordan–Holder theorem isomorphic copies of each must occur in any composition series for U_A. Thus $\dim U \geq t = [E:A]$ and (i) follows.

Lemma 11.5. Let $K[G]$ have a r.b. n and let H be normal in G with $G/H = E$, a group of type (**). Suppose that $|E| > n^3$. Let $G \supseteq Q \supseteq H$ with $Q/H = E'$. Then $[G:Q] \leq n$ and $K[Q]$ has r.b. $[n/2]$.

Proof. By Lemma 10.1(iii) and the fact that $|E'| \neq 0$ in K, we see that $K[E]$ has a nonlinear irreducible module U. Then U is also an irreducible $K[G]$-module and hence by Lemma 11.4

$$[G:Q] = [E:E'] \leq \dim U \leq n.$$

Suppose now that $K[Q]$ does not have r.b. $[n/2]$ and let W be an irreducible $K[Q]$-module with $\dim W > n/2$. Let x be a fixed element of $G - Q$. By Lemma 10.5, $W = V_Q$ for some irreducible $K[G]$-module V and let $\rho : K[G] \to K_d \simeq \operatorname{End}(V)$ be the corresponding representation.

If $y \in Q$, then $x^{-1}yx \in Q$ and we have

$$\begin{aligned}\chi_W(x^{-1}yx) &= \text{trace } \rho(x^{-1}yx) = \text{trace } \rho(x)^{-1}\rho(y)\rho(x)\\ &= \text{trace } \rho(y) = \chi_W(y).\end{aligned}$$

Now let L be a linear module of $K[Q/H]$ viewed as one of $K[Q]$. Then $W \otimes L$ is also irreducible and $\dim W \otimes L = \dim W > n/2$. Thus the above argument applies here and we have

$$\chi_{W\otimes L}(x^{-1}yx) = \chi_{W\otimes L}(y)$$

and hence by Lemma 10.3(ii) and the above

$$\chi_W(y)\chi_L(x^{-1}yx) = \chi_W(x^{-1}yx)\chi_L(x^{-1}yx) = \chi_W(y)\chi_L(y).$$

Since $y \to \chi_L(y)$ is a multiplicative homomorphism we have

$$\chi_W(y) = \chi_W(y)\chi_L(y^{-1}x^{-1}yx).$$

Now let $y \in Q - H$ and use an overbar to denote the natural map $G \to G/H = E$. By Lemma 11.4(ii), we have $\overline{(y^{-1}x^{-1}yx)} = (\bar{y}, \bar{x}) \neq \bar{1}$. Since $K[Q/H]$ is semisimple and commutative by Lemma 10.1(i) and the fact that $|Q/H| = |E'| \neq 0$ in K, we can choose L so that $(\bar{y}, \bar{x})$ does not act like $\bar{1}$ on L. For this L we have $\chi_L(y^{-1}x^{-1}yx) \neq 1$ and hence by the above $\chi_W(y) = 0$. Thus Lemma 10.4(i) applied to the group Q yields

$$|E'| = [Q:H] \leq (\dim W)^2 = (\dim V)^2 \leq n^2.$$

Since $[E:E'] \leq n$, we have $|E| \leq n^3$, a contradiction, and the result follows.

§12. FINITE GROUPS WITH r.b. n

Our main results on finite groups now follow easily. In this section all groups are again assumed to be finite.

Lemma 12.1. Suppose $K[G]$ satisfies a polynomial identity of degree n. Let $\tilde{K}$ be the algebraic closure of K. Then $\tilde{K}[G]$ satisfies a polynomial identity of degree n and all irreducible representations of $\tilde{K}[G]$ have degree less than or equal to $n/2$.

Proof. By Lemma 4.1, $K[G]$ satisfies a multilinear polynomial identity f of degree n. Clearly f is also an identity for $\tilde{K}[G]$. By Theorem 8.1, $\tilde{K}[G]$ has r.b. $[n/2]$.

Lemma 12.2. Let G be a nonabelian group and let K be an algebraically closed field. Suppose that $K[G]$ satisfies a polynomial identity of degree n and that $K[G]$ has r.b. m with $m \leq n/2$. If $|G'| \neq 0$ in K, then G has a normal subgroup $M \neq G$ such that $K[M]$ has r.b. $[m/2]$ and

$$[G:M] \leq (n!)^{2(n!)^2}.$$

Proof. Since G is nonabelian, we can choose H normal in G and maximal subject to the condition that $G/H = E$ is nonabelian. Since $E' = G'H/H$, it follows that

(1) E is nonabelian and $|E'| \neq 0$ in K.

(2) Every proper homomorphic image of E is abelian.

Also $K[E]$ satisfies a polynomial identity of degree n and $K[E]$ has r.b. m with $m \leq n/2$. There are now three cases to consider.

Case (*). E is nilpotent of class 2.

Then E is a type (*) group and we set $M = Z$, the group of Lemma 11.2. Then $[G:M] \leq m^2 \leq (n/2)^2$ by Lemma 11.2 and $K[M]$ has r.b. $[m/2]$.

Case (**). E has a nonidentity normal abelian subgroup, but E is not nilpotent of class 2.

Then E is a type (**) group. If $|E| \leq m^3 \leq (n/2)^3$, then we take $M = H$. By Lemma 10.1(iii) and Lemma 10.5(iii), $K[M]$ has r.b. $[m/2]$. If $|E| > m^3$, then we take $M = Q$, the group of Lemma 11.5. Then $[G:M] \leq m \leq n/2$ and $K[M]$ has r.b. $[m/2]$.

Case (***). E has no nonidentity normal abelian subgroup.

Here we set $M = H$. By Lemmas 10.1(iii) and 10.5(iii), $K[M]$ has r.b. $[m/2]$. It suffices to bound $[G:H] = |E|$. Let N be the normal subgroup of E given by Lemma 9.3 with $[E:N] \leq (n!)^{2(n!)^2}$ and $\mathbf{C}(N) \neq \langle 1 \rangle$. Now $E/\mathbf{C}(N)$ is a proper homomorphic image of E so $\mathbf{C}(N) \supseteq E'$ by (2). If $N \neq \langle 1 \rangle$, then in addition $N \supseteq E'$ so $\mathbf{Z}(N) = N \cap \mathbf{C}(N) \supseteq E'$ and E' is a nonidentity normal abelian subgroup of E, a contradiction. Thus $N = \langle 1 \rangle$ and $|E| = [E:N]$ is suitably bounded. This completes the proof of the lemma.

For convenience we define certain integer valued functions as follows.

$$\mathfrak{A}(n) = (n!)^{n\cdot(n!)^2}$$

$$\mathfrak{A}(n, m) = (n!)^{2m\cdot(n!)^2}.$$

Observe that $\mathfrak{A}(n, 1)$ is the bound given in the previous lemma. We now obtain the first main result of this section.

Theorem 12.3 (Isaacs–Passman [*19*, *20*]). Let G be a finite group and let K be a field with $|G'| \neq 0$ in K. If $K[G]$ satisfies a polynomial identity of degree n, then G has an abelian subgroup A with $[G:A] \leq \mathfrak{A}(n)$.

Proof. By Lemma 12.1 we may assume that K is algebraically closed. We first prove by induction on $|G|$ that if

(1) $|G'| \neq 0$ in K

(2) $K[G]$ satisfies a polynomial identity of degree n

(3) $K[G]$ has r.b. m with $m \leq n/2$

then G has an abelian subgroup A with $[G:A] \leq \mathfrak{A}(n, m)$.

If G is abelian, this is clear since $n, m \geq 1$. Thus let G be nonabelian and let M be the normal subgroup of G given by Lemma 12.2. Then $|M| < |G|$. Since $M \subseteq G$ and $M' \subseteq G'$, it follows that $K[M]$ satisfies (1) and (2) above. By Lemma 12.2, $K[M]$ has r.b. $[m/2]$ and hence r.b. $(m - 1)$ since $m \geq 2$ by Lemma 10.1(iii). Thus by induction M has an abelian subgroup A with $[M:A] \leq \mathfrak{A}(n, m - 1)$. Since $[G:M] \leq \mathfrak{A}(n, 1)$ by Lemma 12.2 we have $[G:A] \leq \mathfrak{A}(n, m - 1)\mathfrak{A}(n, 1) = \mathfrak{A}(n, m)$ and this fact follows.

We can now prove the theorem. Let $K[G]$ satisfy a polynomial identity of degree n so that by Lemma 12.1 $K[G]$ has r.b. $[n/2]$. Thus by the above with $m = [n/2]$ we see that G has an abelian subgroup A with

$$[G:A] \leq \mathfrak{A}(n, [n/2]) \leq \mathfrak{A}(n)$$

and the result follows.

The following theorem is of a more elementary nature. Let H_1 and H_2 be normal solvable subgroups of G. Then $H = H_1H_2$ is normal in G and $H/H_1 \simeq H_2/(H_1 \cap H_2)$ implies easily that H is solvable. It therefore follows that G has a unique maximal solvable normal subgroup which we denote by $\mathbf{S}(G)$.

Theorem 12.4 (Passman [*45*]). Let G be a finite group and suppose that $K[G]$ satisfies a polynomial identity of degree n. Then $[G:\mathbf{S}(G)] \leq \mathfrak{A}(n)$.

Proof. By Lemma 12.1 we may assume that K is algebraically closed. We first prove by induction on $|G|$ that if $K[G]$ satisfies a polynomial identity of degree n and if $K[G]$ has r.b. m, then $[G:\mathbf{S}(G)] \leq \mathfrak{A}(n, m)$. This of course is clear for $|G| = 1$ since $n, m \geq 1$ so we may assume that $G \neq \langle 1 \rangle$.

Suppose first that G has a normal solvable subgroup $H \neq \langle 1 \rangle$. Then $H \subseteq \mathbf{S}(G)$ and if $\bar{G} = G/H$ then clearly $\mathbf{S}(\bar{G}) = \mathbf{S}(G)/H$. Now $K[\bar{G}]$ satisfies a polynomial identity of degree n and $K[\bar{G}]$ has r.b. m and $|\bar{G}| < |G|$. Thus by induction $[\bar{G}:\mathbf{S}(\bar{G})] \leq \mathfrak{A}(n, m)$. Since $[G:\mathbf{S}(G)] = [\bar{G}:\mathbf{S}(\bar{G})]$, the inductive result follows in this case.

Now suppose that G has no normal solvable subgroups $H \neq \langle 1 \rangle$ so that $\mathbf{S}(G) = \langle 1 \rangle$. Let N be the normal subgroup of G given by Lemma 9.3 with $[G:N] \leq \mathfrak{A}(n, 1)$ and with $C = \mathbf{C}(N) \neq \langle 1 \rangle$. Since $\mathbf{S}(G) = \langle 1 \rangle$, we have $N \cap C = \mathbf{Z}(N) = \langle 1 \rangle$ so that G/N contains a subgroup isomorphic to C. Also C is not solvable and hence neither is G/N. By Lemma 10.1(iii), $K[G/N]$ does not have r.b. 1 and hence by Lemma 10.5(iii), $K[N]$ has r.b. $[m/2]$. Now $K[N]$ satisfies a polynomial identity of degree n and $K[N]$ has r.b. $(m - 1)$ since $m \geq 2$ implies that $[m/2] \leq m - 1$. Thus by induction $[N:\mathbf{S}(N)] \leq \mathfrak{A}(n, m - 1)$. Moreover $\mathbf{S}(N)$ is a characteristic subgroup of N and N is normal in G, so $\mathbf{S}(N)$ is a normal solvable subgroup of G and hence $\mathbf{S}(N) = \langle 1 \rangle$. Thus

$$\begin{aligned}[G:\mathbf{S}(G)] &= |N|\,[G:N] \\ &\leq \mathfrak{A}(n, m - 1)\mathfrak{A}(n, 1) = \mathfrak{A}(n, m)\end{aligned}$$

and the inductive result is proved.

We can now prove the theorem. Let $K[G]$ satisfy a polynomial identity of degree n so that by Lemma 12.1, $K[G]$ has r.b. $[n/2]$. Thus by the above with $m = [n/2]$ we see that

$$[G:\mathbf{S}(G)] \leq \mathfrak{A}(n, [n/2]) \leq \mathfrak{A}(n)$$

and the result follows.

§13. SEMIPRIME POLYNOMIAL IDENTITY RINGS

In this section, we obtain a fairly sharp converse to Theorem 5.1. Groups are no longer assumed to be finite.

Lemma 13.1. Let G be a finitely generated group and let m be an integer. Then there exists only finitely many subgroups H of G with $[G:H] \leq m$.

Proof. Let H be a subgroup of G with $[G:H] = t \leq m$. Then G permutes the right cosets of H by right multiplication and this yields a homomorphism $\eta: G \to S_t \subseteq S_m$ where S_m is the symmetric group on m letters. It is clear that the kernel of η is contained in H so that $H = \eta^{-1}(W)$ for some subgroup W of S_m. Now there are only finitely many choices for W and furthermore there are only finitely many η since η is determined by the images of the finite number of generators of G. Thus there are only finitely many possibilities for H.

Lemma 13.2. Let G be an arbitrary group and let m be an integer. Then G has an abelian subgroup with index at most m if and only if every finitely generated subgroup of G has such an abelian subgroup.

Proof. If A is abelian with $[G: A] \leq m$, then for any subgroup H of G we have

$$m \geq [G:A] \geq [G \cap H:A \cap H] = [H:A \cap H].$$

Hence $A \cap H$ is an abelian subgroup of H with index at most m.

Conversely, let us assume that every finitely generated subgroup of G has an abelian subgroup of index at most m. For each finite subset α of G, let $G_\alpha = \langle\alpha\rangle$ be the group generated by the elements in α. Let m_α be the minimum index of abelian subgroups of G_α. By assumption $1 \leq m_\alpha \leq m$ for each α. Choose α_0 such that $m_0 = m_{\alpha_0}$ is the largest of the m_α's and set $G_0 = G_{\alpha_0}$.

Let $A_1, A_2, \ldots, A_r$ be the abelian subgroups of G_0 with $[G_0:A_i] = m_0$. By Lemma 13.1, there are only finitely many of these. We show that for some $i = 1, 2, \ldots, r$ that both $[G:\mathbf{C}(A_i)] \leq m_0$ and $\mathbf{C}(A_i)$ is abelian. This will of course yield the result. Suppose this is not the case. Then for each i, choose α_i to consist of two noncommuting elements of $\mathbf{C}(A_i)$ if the latter is nonabelian or choose α_i to consist of $m_0 + 1$ elements in distinct right cosets of $\mathbf{C}(A_i)$ if $[G:\mathbf{C}(A_i)] > m_0$. Let $\alpha = \alpha_0 \cup \alpha_1 \cup \cdots \cup \alpha_r$. This is a finite set so let A_α be an abelian subgroup of G_α with $[G_\alpha:A_\alpha] = m_\alpha$. Now

$$\begin{aligned} m_0 \geq m_\alpha = [G_\alpha:A_\alpha] &\geq [G_\alpha \cap G_0:A_\alpha \cap G_0] \\ &= [G_0:A_\alpha \cap G_0]. \end{aligned}$$

On the other hand, $A_\alpha \cap G_0$ is an abelian subgroup of G_0 and $m \geq m_0 \geq [G_0:A_\alpha \cap G_0]$ so we must have $[G_0:A_\alpha \cap G_0] = m_0$ by the definition of m_0. Thus $m_0 = m_\alpha$ and $A_\alpha \cap G_0 = A_i$ for some i. Say $A_\alpha \cap G_0 = A_1$.

Since A_α is abelian, we have $A_\alpha \subseteq \mathbf{C}_{G_\alpha}(A_1)$. On the other hand

$$\begin{aligned} [G_\alpha:\mathbf{C}_{G_\alpha}(A_1)] &\geq [G_\alpha \cap G_0:\mathbf{C}_{G_\alpha}(A_1) \cap G_0] \\ &= [G_0:A_1] = m_0 = m_\alpha \end{aligned}$$

since A_1 is clearly its own centralizer in G_0. Thus $A_\alpha = \mathbf{C}_{G_\alpha}(A_1)$. Now $\alpha_1 \subseteq G_\alpha$. Hence if $\mathbf{C}_G(A_1)$ were nonabelian, then α_1 would contain noncommuting elements of $\mathbf{C}_{G_\alpha}(A_1) = A_\alpha$. Since A_α is abelian, this is not the case. On the other hand, if $[G:\mathbf{C}_G(A_1)] > m_0$, then G_α would contain $m_0 + 1$ elements in different right cosets of $\mathbf{C}_G(A_1)$ and hence in different right cosets of $G_\alpha \cap \mathbf{C}_G(A_1) = \mathbf{C}_{G_\alpha}(A_1) = A_\alpha$. But $[G_\alpha : A_\alpha] = m_0$ so we have a contradiction here and the result follows.

Lemma 13.3. Let G be a finitely generated group and let K be any field. Suppose that $K[G]$ satisfies a polynomial identity. Then G is residually finite, that is $\cap N = \langle 1 \rangle$ where N runs over all normal subgroups of G of finite index.

Proof. By Corollary 6.2, G has a normal abelian subgroup A with $[G:A] < \infty$. Moreover A is finitely generated by Lemma 6.1. For each integer m, set $A_m = \{x^m \mid x \in A\}$. Then A_m is a characteristic subgroup of A and hence a normal subgroup of G. Since A is finitely generated, we have $[A:A_m] < \infty$ and $\bigcap_{m=1}^{\infty} A_m = \langle 1 \rangle$.

We now come to our main theorem on group rings satisfying a polynomial identity. The result in characteristic 0 is due to Isaacs and Passman [*19*, 20].

Theorem 13.4 (Passman [44]). Let $K[G]$ be a semiprime group ring which satisfies a polynomial identity of degree n. Then G has an abelian subgroup A with $[G:A] \leq n!\,\mathfrak{A}(n)$.

Proof. Set $m = \mathfrak{A}(n)$. By Theorem 5.5 $[G:\Delta(G)] \leq n!$ and thus it suffices to show that $\Delta = \Delta(G)$ has an abelian subgroup A with $[\Delta:A] \leq m$. Note that since $K[G]$ is semiprime, then either K has characteristic 0 or by Theorem 3.7 K has characteristic $p > 0$ and Δ has no elements of order p.

Suppose by way of contradiction that Δ does not have an abelian subgroup of index $\leq m$. Then by Lemma 13.2, there exists a finitely generated subgroup H of Δ which has no abelian subgroup of index $\leq m$. Now H has only finitely many subgroups of index $\leq m$ by Lemma 13.1 and say these are $L_1, L_2, \ldots, L_t$. By assumption, each is nonabelian so we can choose $x_i \in L_i'$, the commutator subgroup of L_i, with $x_i \neq 1$. Now by Lemma 13.3, H is residually finite and thus for each i we can choose N_i normal in H with $[H:N_i] < \infty$ and $x_i \notin N_i$. Let $N = \cap N_i$. Then N is normal in H, $[H:N] < \infty$ by Lemma 1.1, and $x_i \notin N$ for all i.

By Lemma 6.6, $K[H/N]$ satisfies a polynomial identity of degree n. We consider the finite group $\bar{H} = H/N$. If K has characteristic 0, then certainly $|\bar{H}'| \neq 0$ in K. Suppose K has characteristic $p > 0$. Then by Lemma 2.2, H' is a finite p'-group. Since $\bar{H}' = H'N/N$, we conclude that $|\bar{H}'| \neq 0$ in K in this case also. Thus by Theorem 12.3 $\bar{H}$ has an abelian subgroup $\bar{B}$ of index $\leq m$.

Let B be the complete inverse image of $\bar{B}$ in H. Then $H \supseteq B \supseteq N$ and $B/N = \bar{B}$. Since $[H:B] = [\bar{H}:\bar{B}] \leq m$, we have $B = L_i$ for some i. Thus $L_i/N = B/N$ is abelian and this is a contradiction since $x_i \in L_i'$, $x_i \neq 1$, and $x_i \notin N$. The result follows.

If $K[G]$ is not semiprime, we can still obtain a result of interest. We say that the group H is solvable if it has a finite series of subgroups

$$H_0 = \langle 1 \rangle \subseteq H_1 \subseteq H_2 \subseteq \cdots \subseteq H_t = H$$

such that H_{i-1} is normal in H_i and H_i/H_{i-1} is abelian. It follows easily that solvability is inherited by subgroups and quotient groups. Also if N is a normal solvable subgroup of G with G/N solvable, then G is solvable.

Let G be an arbitrary group and let $\mathscr{S}(G)$ be the set of all solvable normal subgroups of G. If H_1, $H_2 \in \mathscr{S}(G)$, then from $H_1H_2/H_1 \simeq H_2/(H_1 \cap H_2)$ we conclude that $H_1H_2 \in \mathscr{S}(G)$. Hence by induction the same result holds for the subgroup generated by any finite number of $H_i \in \mathscr{S}(G)$. Let

$$\mathbf{S}(G) = \langle H \mid H \in \mathscr{S}(G) \rangle.$$

This is of course consistent with our original definition of $\mathbf{S}(G)$ in the finite case.

Lemma 13.5. $\mathbf{S}(G)$ is a characteristic locally solvable subgroup of G.

Proof. $\mathbf{S}(G)$ is clearly characteristic in G. Let T be a finite subset of $\mathbf{S}(G)$. Then we can find $H_1, H_2, \ldots, H_m \in \mathscr{S}(G)$ with $T \subseteq H_1H_2 \cdots H_m$. Since the latter group is solvable, the lemma is proved.

Theorem 13.6 (Passman [*45*]). Let $K[G]$ satisfy a polynomial identity of degree n. Then $[G:\mathbf{S}(G)] \leq n!\,\mathfrak{A}(n)$.

Proof. We have $[G:\mathbf{S}(G)] \leq [G:\Delta][\Delta:\Delta \cap \mathbf{S}(G)]$. Thus since $[G:\Delta] \leq n!$ by Theorem 5.5, it suffices to show that $[\Delta:\Delta \cap \mathbf{S}(G)] \leq \mathfrak{A}(n)$. Let T be any finite set of coset representatives for $\Delta \cap \mathbf{S}(G)$ in Δ. Since $T \subseteq \Delta$, each element of T has only finitely many conjugates in G and thus if H is the subgroup of G generated by T and all its conjugates, then H is normal in G and H is a finitely generated subgroup of Δ. By Lemma 2.2, $[H:\mathbf{Z}(H)] < \infty$ and we let W be the subgroup of H given by $H \supseteq W \supseteq \mathbf{Z}(H)$ and $W/\mathbf{Z}(H) = \mathbf{S}(H/\mathbf{Z}(H))$. Clearly W is a normal solvable subgroup of G so that $W \subseteq \mathbf{S}(G)$.

Now $K[G]$ satisfies a polynomial identity of degree n and hence so does $K[H/\mathbf{Z}(H)]$. Thus by Theorem 12.4 we have

$$[H:W] = [H/\mathbf{Z}(H):\mathbf{S}(H/\mathbf{Z}(H))] \leq \mathfrak{A}(n).$$

Since $T \subseteq H$ and the elements of T are in distinct cosets of $\Delta \cap \mathbf{S}(G) \supseteq W$, we have $|T| \leq \mathfrak{A}(n)$ and the result follows.

Corollary 13.7. Let G be a finitely generated group and let $K[G]$ satisfy a polynomial identity of degree n. Then $\mathbf{S}(G)$ is solvable and $[G:\mathbf{S}(G)] \leq \mathfrak{A}(n)$.

Proof. By Corollary 6.2, G has a normal abelian subgroup A of finite index. Let S be the subgroup of G given by $G \supseteq S \supseteq A$ and $S/A = \mathbf{S}(G/A)$. Clearly S is a normal solvable subgroup of G, so $S \subseteq \mathbf{S}(G)$. On the other hand if $H \in \mathscr{S}(G)$, then $HA/A \in \mathscr{S}(G/A)$ so $H \subseteq S$. Thus $S = \mathbf{S}(G)$ is solvable.

Now $K[G]$ satisfies a polynomial identity of degree n and hence so does $K[G/A]$. Thus by Theorem 12.4 we have $[G:S] = [G/A:\mathbf{S}(G/A)] \leq \mathfrak{A}(n)$ and the result follows.

§14. EXAMPLES

We now consider some examples to show that an exact converse to Theorem 5.1 is false in general. Recall that if R is a ring and if R_1 and R_2 are subsets of R, then $[R_1, R_2]$ is the set of all finite sums of Lie products $[\alpha_1, \alpha_2] = \alpha_1\alpha_2 - \alpha_2\alpha_1$ with $\alpha_i \in R_i$. We define $\gamma^n R$ inductively by $\gamma^0 R = R$, $\gamma^{n+1}R = [\gamma^n R, R]$. Then R is said to be Lie nilpotent if $\gamma^n R = 0$ for some n and the class of R is given by the minimal such n.

Lemma 14.1. Let E be an algebra over K and suppose that $[E, E]^n = 0$. Then E satisfies the standard polynomial identity of degree $2n$.

Proof. Let $\alpha_1, \alpha_2, \ldots, \alpha_{2n} \in E$ and consider

$$[\alpha_1, \alpha_2, \ldots, \alpha_{2n}] = \sum_{\sigma} (-1)^{\sigma} \alpha_{\sigma(1)}\alpha_{\sigma(2)} \cdots \alpha_{\sigma(2n)}.$$

Consider all such terms on the right-hand side with $\{\sigma(1), \sigma(2)\} = \{i_1, i_2\}$, $\{\sigma(3), \sigma(4)\} = \{i_3, i_4\}, \ldots, \{\sigma(2n-1), \sigma(2n)\} = \{i_{2n-1}, i_{2n}\}$, where of course $\{i_1, i_2, \ldots, i_{2n}\} = \{1, 2, \ldots, 2n\}$. Then the subsum $\sum'$ of all these terms is easily seen to be

$$\textstyle\sum' = \pm[\alpha_{i_1}, \alpha_{i_2}][\alpha_{i_3}, \alpha_{i_4}] \cdots [\alpha_{i_{2n-1}}, \alpha_{i_{2n}}] = 0$$

since $[E, E]^n = 0$. The result clearly follows.

Lemma 14.2. Let K be a field of characteristic $p > 0$ and let G be a group with $|G'| = p$ and G' central in G. Then

(i) $K[G]$ satisfies the standard polynomial identity of degree $2p$.

(ii) $K[G]$ is Lie nilpotent of class at most p and hence satisfies a polynomial identity of degree $p + 1$.

Proof. Since $|G'| = p$, $G' = \langle z \rangle$ is cyclic. We show first that $[K[G], K[G]] \subseteq (1 - z)K[G]$. Now $[K[G], K[G]]$ is spanned over K by elements of the form $[x, y]$ with $x, y \in G$. For $x, y \in G$, we have

$$\begin{aligned}[x, y] &= xy - yx = (1 - yxy^{-1}x^{-1})xy \\ &= (1 - z^i)xy = (1 - z)(1 + z + \cdots + z^{i-1})xy\end{aligned}$$

for some $i > 0$ since $yxy^{-1}x^{-1} \in G' = \langle z \rangle$. Thus $[x, y] \in (1 - z)K[G]$ and this fact follows.

Now K has characteristic p and $z^p = 1$ so $(1 - z)^p = 1 - z^p = 0$. Since z is central in G, we have $((1 - z)K[G])^p = 0$ and (i) follows from Lemma 14.1. In addition we have easily by induction $\gamma^n K[G] \subseteq (1 - z)^n K[G]$ and thus $\gamma^p K[G] = 0$. Hence $K[G]$ is Lie nilpotent of class at most p and $K[G]$ satisfies the polynomial identity

$$f(\zeta_0, \zeta_1, \ldots, \zeta_p) = [[\cdots[[\zeta_0, \zeta_1]\zeta_2]\cdots]\zeta_p]$$

of degree $p + 1$.

Theorem 14.3. Let K be a field of characteristic $p > 0$. Then there exists a sequence of finite p-groups $P_1, P_2, \ldots, P_n, \ldots$ and an infinite p-group P_∞ such that

(i) For all $\nu = 1, 2, \ldots, \infty$, $K[P_\nu]$ satisfies the standard polynomial identity of degree $2p$.

(ii) P_n has no abelian subgroup of index less than p^n.

(iii) P_∞ has no abelian subgroup of finite index.

Proof. Let Q be a nonabelian group of order p^3. Then Z, the center of Q, has order p, Q/Z is abelian of type (p, p), and $Q' = Z$. Let $Q_1, Q_2, Q_3, \ldots$ be copies of Q with centers $Z_1, Z_2, Z_3, \ldots$ respectively and say $Z_i = \langle z_i \rangle$. For each integer n set

$$G_n = Q_1 \times Q_2 \times \cdots \times Q_n$$

and set

$$G_\infty = Q_1 \times Q_2 \times \cdots \times Q_n \times \cdots.$$

We have clearly $G'_\nu = \mathbf{Z}(G_\nu) = Z_1 \times Z_2 \times \cdots$. Now let N_ν be the subgroup of $\mathbf{Z}(G_\nu)$ generated by the elements $z_2 z_1^{-1}, z_3 z_1^{-1}, z_4 z_1^{-1}, \ldots$. Then N_ν is a central and hence a normal subgroup of G_ν and we set

$$P_n = G_n/N_n, \qquad P_\infty = G_\infty/N_\infty.$$

Clearly $P'_\nu \subseteq \mathbf{Z}(G_\nu)/N_\nu$ and the latter group has order p. Thus $|P'_\nu| \leq p$ and P'_ν is central so (i) follows by Lemma 14.2. We observe now that $\mathbf{Z}(P_\nu) = \mathbf{Z}(G_\nu)/N_\nu$. For suppose $x = x_1 x_2 \cdots \in G_\nu - \mathbf{Z}(G_\nu)$ with $x_i \in Q_i$.

Then for some i, $x_i \notin Z_i$ and hence there exists $y_i \in Q_i$ which does not centralize x_i. Then $y_i \in G_\nu$ and

$$(x, y_i) = x^{-1}y_i^{-1}xy_i = x_i^{-1}y_i^{-1}x_iy_i$$

is a nonidentity element of Z_i. Since clearly $Z_i \cap N_\nu = \langle 1 \rangle$, we see that the images of x and y_i do not commute in P_ν. This yields $[P_n : \mathbf{Z}(P_n)] = p^{2n}$ and $[P_\infty : \mathbf{Z}(P_\infty)] = \infty$.

Suppose A is an abelian subgroup of P_ν of finite index $\leq p^t$ and set $B = A\mathbf{Z}(P_\nu)$. Then B is abelian of index $\leq p^t$ and B is normal in P_ν since $B \supseteq \mathbf{Z}(P_\nu) = P_\nu'$. Now P_ν/B is clearly elementary abelian and we can choose $w_1, w_2, \ldots, w_t \in P_\nu$ with $P_\nu = \langle B, w_1, w_2, \ldots, w_t \rangle$. If $y \in P_\nu$, then $y^{-1}w_iy = w_i(w_i, y) \in w_iP_\nu'$. Hence since $|P_\nu'| = p$, we see that w_i has at most p conjugates in P_ν and $[P_\nu : \mathbf{C}_{P_\nu}(w_i)] \leq p$. Thus by Lemma 1.1 if

$$W = B \cap \mathbf{C}_{P_\nu}(w_1) \cap \mathbf{C}_{P_\nu}(w_2) \cap \cdots \cap \mathbf{C}_{P_\nu}(w_t)$$

then $[P_\nu : W] \leq p^t \cdot p \cdot p \cdots p = p^{2t}$. Now B is abelian so W centralizes B and all the w_i and hence $W = \mathbf{Z}(P_\nu)$. Since $[P_\infty : \mathbf{Z}(P_\infty)] = \infty$, (iii) follows and since $[P_n : \mathbf{Z}(P_n)] = p^{2n}$ we have $t \geq n$ and (ii) follows. This completes the proof.

The above finite examples show that the assumption that $|G'| \neq 0$ in K is in fact a necessary ingredient of Theorem 12.3. In addition, the example of the group P_∞ shows that the assumption that G is finitely generated is necessary for Corollary 6.2. Thus there is still much work to be done on polynomial identities in case $K[G]$ is not semiprime and the obvious first step must certainly be the study of finite p-groups in characteristic p.

Let P be a nonabelian p-group of order p^3 and let K be a field of characteristic p. By Lemma 14.2, $K[P]$ satisfies a polynomial identity of degree $p + 1$. Now it would be interesting to find the minimal degree n of the polynomial identities satisfied by $K[P]$ and in fact by Lemma 9.4 we have $(n!)^2 \geq p$. A somewhat better bound is given below.

Theorem 14.4. Let K be a field and let P be a finite nonabelian p group. If $K[P]$ satisfies a polynomial identity of degree n, then $n > (6p)^{\frac{1}{3}}$.

Proof. Since $P' \neq \langle 1 \rangle$, we can choose N normal in P with $[P' : N] = p$. If $\bar{P} = P/N$, then $K[\bar{P}]$ also satisfies a polynomial identity of degree n and $\bar{P}'$ is central of order p. Thus it suffices to assume that $P = \bar{P}$ or equivalently that P' is central of order p. Choose $x, y \in P$ which do not commute and let $z = xyx^{-1}y^{-1} \neq 1$. Then $P' = \langle z \rangle$ and z is central of order p. We have $xy = zyx$ and hence by induction $xy^i = z^iy^ix$.

We may assume by Lemma 4.1 that $K[P]$ satisfies the multilinear identity

$$f(\zeta_1, \zeta_2, \ldots, \zeta_n) = \zeta_1\zeta_2 \cdots \zeta_n + \sum_{\sigma \neq 1} a_\sigma \zeta_{\sigma(1)}\zeta_{\sigma(2)} \cdots \zeta_{\sigma(n)}$$

and we set $\zeta_i = xy^i$. Consider the σ monomial evaluated here. We have

$$\mu_\sigma = xy^{\sigma(1)}xy^{\sigma(2)} \cdots xy^{\sigma(n)}.$$

Now shift all the y's to the left starting with $y^{\sigma(1)}$. Since $y^{\sigma(i)}$ must pass precisely i of the x's we have using $xy^j = z^jy^jx$

$$\begin{aligned}\mu_\sigma &= z^{\sum_1^n i\sigma(i)} y^{\sum_1^n \sigma(i)} x^n \\ &= z^{\sum_1^n i\sigma(i)} y^{n(n+1)/2} x^n.\end{aligned}$$

For each σ we define the integral vector v_σ by

$$v_\sigma = \langle\langle \sigma(1), \sigma(2), \ldots, \sigma(n) \rangle\rangle$$

so that $v_1 = \langle\langle 1, 2, \ldots, n \rangle\rangle$ and

$$\mu_\sigma = z^{v_1 \cdot v_\sigma} y^{n(n+1)/2} x^n$$

where $v_1 \cdot v_\sigma$ is the usual dot product of vectors.

We consider the values $v_1 \cdot v_\sigma$. By the Cauchy–Schwarz inequality

$$(v_1 \cdot v_\sigma)^2 \leq (v_1 \cdot v_1)(v_\sigma \cdot v_\sigma)$$

with equality if and only if v_σ is a scalar multiple of v_1 and hence in our case if and only if $\sigma = 1$. Since clearly $v_\sigma \cdot v_\sigma = v_1 \cdot v_1$, we have for all $\sigma \neq 1$, $(v_1 \cdot v_\sigma)^2 < (v_1 \cdot v_1)^2$ and thus

$$v_1 \cdot v_\sigma < v_1 \cdot v_1 = \sum_1^n i^2 = n(n + 1)(2n + 1)/6$$

since these are all positive quantities.

Clearly zero is a lower bound for all $v_1 \cdot v_\sigma$. A better bound is obtained as follows. For each σ define σ' by $\sigma'(i) = n + 1 - \sigma(i)$. Thus σ' is clearly also a permutation and $v_{\sigma'} = v - v_\sigma$ where

$$v = \langle\langle n + 1, n + 1, \ldots, n + 1 \rangle\rangle.$$

By the above

$$v_1 \cdot v_1 \geq v_1 \cdot v_{\sigma'} = v_1 \cdot (v - v_\sigma) = v_1 \cdot v - v_1 \cdot v_\sigma$$

and thus

$$\begin{aligned}v_1 \cdot v_\sigma \geq v_1 \cdot v - v_1 \cdot v_1 &= (n + 1) \sum_1^n i - \sum_1^n i^2 \\ &= n(n + 1)(n + 2)/6.\end{aligned}$$

Returning to the polynomial identity, we have

$$\begin{aligned} 0 &= f(xy^1, xy^2, \ldots, xy^n) \\ &= \left(z^{v_1 \cdot v_1} + \sum_{\sigma \neq 1} a_\sigma z^{v_1 \cdot v_\sigma}\right) y^{n(n+1)/2} x^n. \end{aligned}$$

It therefore follows that the $z^{v_1 \cdot v_1}$ term must be canceled by other terms in the sum and thus for some $\sigma \neq 1$ we have $z^{v_1 \cdot v_1} = z^{v_1 \cdot v_\sigma}$. Hence $p \mid (v_1 \cdot v_1 - v_1 \cdot v_\sigma)$. Now by the above

$$\begin{aligned} 0 < v_1 \cdot v_1 - v_1 \cdot v_\sigma &\leq n(n+1)(2n+1)/6 - n(n+1)(n+2)/6 \\ &= n(n-1)(n+1)/6 \end{aligned}$$

and since p divides $v_1 \cdot v_1 - v_1 \cdot v_\sigma$ we must have $n(n-1)(n+1)/6 \geq p$. Thus $n^3 > n(n^2 - 1) \geq 6p$ and $n > (6p)^{\frac{1}{3}}$. This completes the proof.

III

NIL AND NILPOTENT IDEALS

§15. RELATIVE PROJECTIVITY

Let S be a ring. If V is a right S-module, then we will write $V = V_S$ to stress its dependence on the ring. Let W be a submodule of V. Then we write $W_S \mid V_S$ if there exists an S-submodule U of V with $V_S = W_S + U_S$, a direct sum. Now let R be a subring of S with the same 1. Then by restriction any S-module V_S can be viewed as an R-module V_R. We say that S is R-projective if given any S-module V_S and submodule W_S, then $W_R \mid V_R$ implies $W_S \mid V_S$. In this section, we offer three pairs (S, R) with S an R-projective overring.

Lemma 15.1. Let R be a ring with 1 and let $S = R_n$. Then S is R-projective where R is contained in S as the set of scalar matrices.

Proof. Let $\{e_{ij}\}$ be a set of matrix units in S.

Let V be an S-module and let W be a submodule and suppose that $W_R \mid V_R$. This implies that there exists a projection map $f\colon V \to W$ such that $f(w) = w$ for all $w \in W$ and $f(v\alpha) = f(v)\alpha$ for all $v \in V$, $\alpha \in R$. We define $g\colon V \to W$ by

$$g(v) = \sum_{i=1}^{n} f(ve_{i1})e_{1i}$$

and this makes sense since $f(ve_{i1}) \in W$ and W is an S-submodule of V. If $\alpha \in R$, then since α commutes with all e_{ij} we have clearly $g(v\alpha) = g(v)\alpha$ for all $v \in V$. Also

$$g(ve_{jk}) = \sum_{i=1}^{n} f(ve_{jk}e_{i1})e_{1i} = f(ve_{j1})e_{1k}$$

$$g(v)e_{jk} = \sum_{i=1}^{n} f(ve_{i1})e_{1i}e_{jk} = f(ve_{j1})e_{1k}$$

so $g(v\alpha) = g(v)\alpha$ for all $\alpha \in S$. If $w \in W$, then $we_{i1} \in W$ so $f(we_{i1}) = we_{i1}$ and

$$g(w) = w \sum_{1}^{n} e_{i1}e_{1i} = w.$$

Set $U = \{v \in V \mid g(v) = 0\}$. It follows easily that U is an S-submodule of V and $V = W + U$ is a direct sum. The result follows.

The next lemma is due to D. G. Higman [*17*].

Lemma 15.2. Let K be a field, G a group, and H a subgroup of G of finite index. If $[G:H] = n$ is not zero in K, then $K[G]$ is $K[H]$-projective.

Proof. Let $x_1, x_2, \ldots, x_n$ be a set of right coset representatives for H in G. By assumption $1/n \in K$.

Let V be a $K[G]$-module and let W be a submodule and suppose that $W_R \mid V_R$ where $R = K[H]$. This implies that there exists a projection map $f: V \to W$ such that $f(w) = w$ for all $w \in W$ and $f(v\alpha) = f(v)\alpha$ for all $v \in V$, $\alpha \in R$. We define $g: V \to W$ by

$$g(v) = \frac{1}{n}\sum_{i=1}^{n} f(vx_i^{-1})x_i$$

and this makes sense since $f(vx_i^{-1}) \in W$ and W is a $K[G]$-submodule. Let $x \in G$. Then x permutes the right cosets $Hx_1, Hx_2, \ldots, Hx_n$ of H by right multiplication and thus $x_ix = h_ix_{i'}$ where $h_i \in H$ and $i \to i'$ is a permutation of $\{1, 2, \ldots, n\}$. Since $x_i^{-1}h_i = xx_{i'}^{-1}$ we then have

$$\begin{aligned} g(v)x &= \frac{1}{n}\sum_{i=1}^{n} f(vx_i^{-1})x_ix = \frac{1}{n}\sum_{i=1}^{n} f(vx_i^{-1})h_ix_{i'} \\ &= \frac{1}{n}\sum_{i=1}^{n} f(vx_i^{-1}h_i)x_{i'} \\ &= \frac{1}{n}\sum_{i=1}^{n} f(vxx_{i'}^{-1})x_{i'} = g(vx). \end{aligned}$$

Thus clearly $g(v\alpha) = g(v)\alpha$ for all $\alpha \in K[G]$. If $w \in W$, then $wx_i^{-1} \in W$ so $f(wx_i^{-1}) = wx_i^{-1}$ and

$$g(w) = \frac{1}{n}\sum_1^n wx_i^{-1}x_i = w.$$

Set $U = \{v \in V \mid g(v) = 0\}$. It follows easily that U is a $K[G]$-submodule of V and that $V = W + U$ is a direct sum. This completes the proof.

Theorem 15.3 (Maschke). Let G be a finite group and let K be a field with $|G| \neq 0$ in K. Then every finite dimensional $K[G]$-module is a direct sum of irreducible $K[G]$-modules.

Proof. By Lemma 15.2 with $H = \langle 1 \rangle$ we see that $K[G]$ is K-projective. Let V be a $K[G]$-module and let W be a submodule. Since K is a field, it is clear that $W_K \mid V_K$ and hence $W \mid V$. The result now follows by induction on the dimension of V.

Let A be an algebra over a field K and let F be an extension field of K. Then we let A^F denote the F-algebra $F \otimes_K A$. We note that A is naturally embedded in A^F by way of the injective map $\alpha \to 1 \otimes \alpha$ for $\alpha \in A$ and similarly $F \subseteq A^F$ by way of the map $a \to a \otimes 1$ for $a \in F$. Moreover with these embeddings we have $A^F = F \cdot A$.

Lemma 15.4. Let A be an algebra over a field K and let F be a finite separable field extension of K. Then $A^F = F \otimes_K A$ is A-projective.

Proof. Let $(F:K) = n$. Since F/K is separable, it follows that $F = K[\pi]$ for some primitive element $\pi \in F$ and that $\pi^n = c_0 + c_1\pi + \cdots + c_{n-1}\pi^{n-1}$ for some $c_0, c_1, \ldots, c_{n-1} \in K$. Furthermore the above polynomial equation is irreducible and separable so that π is not a root of the derivative polynomial. Hence $n\pi^{n-1} \neq \sum_0^{n-1} ic_i\pi^{i-1}$ so

$$n\sum_0^{n-1} c_i\pi^i = n\pi^n \neq \sum_0^{n-1} ic_i\pi^i$$

and $\sum_0^{n-1} (n-i)c_i\pi^i = d \neq 0$. Let F and A be embedded naturally in A^F. Then $\{1, \pi, \pi^2, \ldots, \pi^{n-1}\}$ is a basis for A^F over A.

Let V be an A^F-module and let W be a submodule and suppose that $W_A \mid V_A$. This implies that there exists a projection map $f: V \to W$ such that $f(w) = w$ for all $w \in W$ and $f(v\alpha) = f(v)\alpha$ for all $\alpha \in A$. Define $g: V \to W$ by

$$g(v) = \sum_{i=0}^{n-1} f(v\pi^i)\pi^{-i}(c_0 + c_1\pi + \cdots + c_i\pi^i)d^{-1}$$

and this makes sense since $f(v\pi^i) \in W$ and W is an A^F-submodule. If $\alpha \in A$, then since α commutes with all elements of F we have clearly $g(v\alpha) = g(v)\alpha$. We now consider $g(v\pi)$. We have

$$\begin{aligned}
g(v\pi) &= \sum_0^{n-1} f(v\pi^{i+1})\pi^{-i}(c_0 + c_1\pi + \cdots + c_i\pi^i)d^{-1} \\
&= f(v\pi^n)\pi d^{-1} + \sum_0^{n-2} f(v\pi^{i+1})\pi^{-i}(c_0 + c_1\pi + \cdots + c_i\pi^i)d^{-1} \\
&= \sum_0^{n-1} f(v\pi^i)c_i\pi d^{-1} + \sum_1^{n-1} f(v\pi^i)\pi^{-i+1}(c_0 + c_1\pi + \cdots + c_{i-1}\pi^{i-1})d^{-1} \\
&= \sum_0^{n-1} f(v\pi^i)\pi^{-i+1}(c_0 + c_1\pi + \cdots + c_i\pi^i)d^{-1} \\
&= g(v)\pi
\end{aligned}$$

and thus by induction $g(v\pi^j) = g(v)\pi^j$ for all $j = 0, 1, \ldots, n-1$. This implies that $g(v\beta) = g(v)\beta$ for all $\beta \in A^F$. If $w \in W$, then $w\pi^i \in W$ so $f(w\pi^i) = w\pi^i$ and

$$\begin{aligned}
g(w) &= wd^{-1}\sum_0^{n-1}(c_0 + c_1\pi + \cdots + c_i\pi^i) \\
&= wd^{-1}\sum_0^{n-1}(n - i)c_i\pi^i \\
&= wd^{-1}d = w.
\end{aligned}$$

Set $U = \{v \in V \mid g(v) = 0\}$. It follows easily that U is an A^F-submodule of V and that $V = W + U$ is a direct sum. This completes the proof.

§16. FINITE EXTENSIONS

There are a number of interesting applications of relative projectivity to the study of the relationship between the Jacobson radicals of a ring S and its subring R. Recall first that JS is the intersection of all primitive ideals in S. An alternate characterization of JS is as follows. An ideal I of S is said to be quasiregular if for each $\alpha \in I$ there exists $\beta \in S$ with $\alpha + \beta + \alpha\beta = 0$. Then as is well known, JS is the unique quasiregular ideal which contains all the quasiregular ideals of S. We note that the equation $\alpha + \beta + \alpha\beta = 0$ is equivalent to $(1 + \alpha)(1 + \beta) = 1$. Since any nil ideal is quasiregular, it follows that JS contains all nil ideals of S. The next result is Nakayama's lemma.

Lemma 16.1. Let S be a ring and let V be a finitely generated S-module. If $V = V(JS)$, then $V = 0$.

Proof. Suppose V has n generators $v_1, v_2, \ldots, v_n$. Then $v_n \in V = V(JS)$ so we can write $v_n = v_1\alpha_1 + v_2\alpha_2 + \cdots + v_n\alpha_n$ with $\alpha_i \in JS$. Hence $-\alpha_n \in JS$ and $1 - \alpha_n$ has a right inverse $\gamma \in S$. This yields $v_n(1 - \alpha_n) = \sum_1^{n-1} v_i\alpha_i$ and $v_n = \sum_1^{n-1} v_i(\alpha_i\gamma)$. It now follows that V is generated by $v_1, v_2, \ldots, v_{n-1}$. We continue this process until all generators of V have been eliminated and thus $V = 0$.

Lemma 16.2. Let S be a ring and let V be an S-module. Suppose that $V = V_1 + V_2 + \cdots + V_n$ is a finite direct sum of the irreducible S-modules V_i. If $U \subseteq W$ are submodules of V, then $U \mid W$.

Proof. For each subset $\tau \subseteq \{1, 2, \ldots, n\}$, let $V_\tau = \sum_{i\in\tau} V_i$, $V_{\tau'} = \sum_{i\notin\tau} V_i$ so that $V = V_\tau + V_{\tau'}$ is a direct sum. Let U be a submodule of V and let $\tau \subseteq \{1, 2, \ldots, n\}$ be maximal subject to $U \cap V_\tau = 0$. Then $U + V_\tau$ is a direct sum. Suppose $U + V_\tau \neq V$. Then for some j, $U + V_\tau \not\supseteq V_j$ so since V_j is an irreducible S-module we have $(U + V_\tau) \cap V_j = 0$ and thus $U \cap (V_\tau + V_j) = 0$, a contradiction. We have therefore shown that $U + V_\tau = V$ and hence $U \mid V$. In addition since $V_{\tau'} + V_\tau = V$ we have $U \simeq V/V_\tau \simeq V_{\tau'}$ so U is a finite direct sum of irreducible S-modules.

Now let $U \subseteq W \subseteq V$ be given. Then by the above applied to $W \subseteq V$ we see that W is a finite direct sum of irreducible S-modules. Then by the above applied to $U \subseteq W$ we have $U \mid W$.

Now let S be a ring and let R be a subring with the same 1. We say that $\{x_1 = 1, x_2, \ldots, x_n\}$ is a normalizing basis for S over R if

(1) Every element of S can be written uniquely as

$$\alpha = \beta_1 x_1 + \beta_2 x_2 + \cdots + \beta_n x_n$$

with $\beta_i \in R$.

(2) There exist automorphisms $\sigma_1, \sigma_2, \ldots, \sigma_n$ of R such that $x_i\beta = \sigma_i(\beta)x_i$ for all $\beta \in R$. Since $x_1 = 1$, we have of course $\sigma_1 = 1$, the identity automorphism.

Theorem 16.3. Let S be a ring and let R be a subring with the same 1. If $\{x_1 = 1, x_2, \ldots, x_n\}$ is a normalizing basis for S over R, then

$$(JS)^n \subseteq (JR) \cdot S \subseteq JS.$$

Furthermore, if S is R-projective, then $JS = (JR) \cdot S$.

Proof. We show first that $JR \subseteq JS$. Let V be an irreducible S-module and fix $v \in V - \{0\}$ so that $V = vS$. By (1) and (2) above it follows that

$$V = vx_1R + vx_2R + \cdots + vx_nR$$

so that V_R is a finitely generated R-module. Set $W = V(JR)$. Then for any $i = 1, 2, \ldots, n$ and any $\beta \in R$, we have by (2)

$$\begin{aligned} W\beta x_i &= V(JR)x_i = Vx_i\sigma_i^{-1}(JR) \\ &\subseteq V(JR) = W \end{aligned}$$

since σ_i^{-1} being an automorphism of R implies that $\sigma_i^{-1}(JR) = JR$. Thus by (1), W is an S-submodule of V and hence since V is irreducible we have either $W = V$ or $W = 0$. If $W = V$, then $V = V(JR)$ and since V_R is a finitely generated R-module and $V \neq 0$ this contradicts Lemma 16.1. Thus $0 = W = V(JR)$. We have shown that JR acts trivially on every irreducible S-module and hence $JR \subseteq JS$. Since JS is an ideal, we have $(JR)S \subseteq JS$.

We now consider the reverse inclusions. This will require inducing representations from R to S and we take a rather concrete approach. First by (1), $S = X$ is a free left R-module of rank n and also a right S-module and these actions commute. Since S acts faithfully on X, we therefore have an injective homomorphism of S into R_n given by $\alpha \to (\alpha_{ij}) \in R_n$ where $x_i\alpha = \sum \alpha_{ij}x_j$ and the $\alpha_{ij} \in R$ are uniquely determined by (1). We remark that if $\beta \in R$, then $x_i\beta = \sigma_i(\beta)x_i$ so β maps to the diagonal matrix $\operatorname{diag}(\sigma_1(\beta), \sigma_2(\beta), \ldots, \sigma_n(\beta))$. If V is an R-module, then we define the induced S-module $\bar{V}$ as follows. $\bar{V}$ is the set of all n-tuples $\bar{v} = (v_1, v_2, \ldots, v_n)$ with entries in V and S acts on $\bar{V}$ by way of the matrix product $\bar{v}\alpha = (v_1, v_2, \ldots, v_n)(\alpha_{ij})$. Let V_i be the subset of $\bar{V}$ consisting of those n-tuples which are zero except in the ith position. Since R acts diagonally on $\bar{V}$, it follows that each V_i is an R-submodule of $\bar{V}_R$ and in fact $\bar{V}_R = V_1 + V_2 + \cdots + V_n$ is a direct sum. We remark that V_1 is R-isomorphic to V since $\sigma_1 = 1$ and that if V is an irreducible R-module, then so is each V_i since σ_i is an automorphism of R. Furthermore since $x_1 = 1$ by (1), we have $x_1x_i = 1 \cdot x_i$ and this implies that the map $v_1 \to v_1x_i$ is a one-to-one map of V_1 onto V_i.

Now let V be an irreducible R-module and consider the induced S-module $\bar{V}$. Then by the above, $\bar{V}_R$ is a direct sum of n irreducible R-modules. Hence clearly $\bar{V}$ has a composition series of length less than or equal to n. Since JS acts trivially on each factor of this series, we see that $(JS)^n$ acts trivially on $\bar{V}$. Let $\alpha \in (JS)^n$ with $\alpha = \sum \alpha_ix_i$, $\alpha_i \in R$ by (1) above. Then $\bar{V}\alpha = 0$ so we have using $\sigma_1 = 1$

$$0 = v_1\alpha = \sum_i (v_1\alpha_i)x_i$$

for all $v_1 \in V_1$. Since $v_1\alpha_i \in V_1$ and since $v_1 \to v_1x_i$ is a one-to-one map of V_1 onto V_i, this then yields $v_1\alpha_i = 0$. Now V_1 is R-isomorphic to V and thus $V\alpha_i = 0$ for $i = 1, 2, \ldots, n$. Since this is true for all such irreducible R-modules V, we conclude that $\alpha_i \in JR$. Thus $\alpha \in (JR) \cdot S$ and hence $(JS)^n \subseteq (JR) \cdot S$.

Finally suppose that in addition S is R-projective. Let V and $\bar{V}$ be as above. Since $\bar{V}_R$ is a finite direct sum of irreducible R-modules and since S is R-projective, it follows easily from Lemma 16.2 that $\bar{V}$ is a finite direct sum of irreducible S-modules. Thus $\bar{V}(JS) = 0$. Then as in the preceding paragraph $\bar{V}\alpha = 0$ for all such V implies that $\alpha \in (JR) \cdot S$ and hence $JS \subseteq (JR) \cdot S$. Since we have previously shown that $JS \supseteq (JR) \cdot S$, we have $JS = (JR) \cdot S$ and the result follows.

We can now easily obtain a number of corollaries.

Theorem 16.4. Let R be a ring. Then $J(R_n) = (JR)_n$.

Proof. If $S = R_n$, then by Lemma 15.1, S is R-projective. Clearly $\{1\} \cup \{e_{ij} \mid (i,j) \neq (1, 1)\}$ is a normalizing basis for S over R with all σ's equal to 1. Thus by Theorem 16.3, $J(R_n) = (JR) \cdot R_n$ and the latter is clearly equal to $(JR)_n$.

Lemma 16.5. Let S be a ring and let R be a subring with the same 1. Suppose that as left R-modules we have $R_R \mid S_R$. Then $JS \cap R \subseteq JR$.

Proof. Clearly $JS \cap R$ is an ideal of R and it suffices to show that it is quasiregular. Let $\alpha \in JS \cap R$. Then $\alpha \in JS$ and hence there exists $\beta \in S$ with $\alpha + \beta + \alpha\beta = 0$. Now $R_R \mid S_R$ as left R-modules so say $S = R + U$, a direct sum, with $RU \subseteq U$. Then we can write $\beta = \beta_0 + \beta_1$ with $\beta_0 \in R$, $\beta_1 \in U$, and thus

$$0 = \alpha + \beta + \alpha\beta = (\alpha + \beta_0 + \alpha\beta_0) + (\beta_1 + \alpha\beta_1).$$

Now $\alpha + \beta_0 + \alpha\beta_0 \in R$, $\beta_1 + \alpha\beta_1 \in U$, and $R \cap U = 0$ so we have $\alpha + \beta_0 + \alpha\beta_0 = 0$. Since $\beta_0 \in R$, this shows that $JS \cap R$ is a quasiregular ideal of R which is therefore contained in JR.

The following few results are basic to the study of the relationship between $JK[H]$ and $JK[G]$ where H is a subgroup of G.

Theorem 16.6. Let H be a normal subgroup of G with $[G:H] = n < \infty$. Then

$$(JK[G])^n \subseteq (JK[H]) \cdot K[G] \subseteq JK[G].$$

If in addition $n \neq 0$ in K, then $JK[G] = (JK[H]) \cdot K[G]$.

Proof. Let $x_1 = 1, x_2, \ldots, x_n$ be a set of coset representatives for H in G. Since H is normal in G, these clearly form a normalizing basis for $K[G]$ over $K[H]$. The result follows from Lemma 15.2 and Theorem 16.3.

Lemma 16.7. Let H be a normal subgroup of G. Then

$$(JK[H]) \cdot K[G] = K[G] \cdot (JK[H]).$$

Proof. Since H is normal in G, conjugation by $x \in G$ induces an automorphism of $K[H]$. Thus $x^{-1}JK[H]x = JK[H]$ and the result follows.

Lemma 16.8. Let $[G:H] < \infty$. Then $JK[G]$ is nilpotent if and only if $JK[H]$ is nilpotent.

Proof. If H is normal in G, the result follows from Theorem 16.6 and Lemma 16.7. In general let H_0 be the core of H, that is H_0 is the intersection of the at most $[G:H]$ conjugates of H in G. Then H_0 is normal in G and $[G:H_0] < \infty$ by Lemma 1.1. Thus $JK[G]$ is nilpotent if and only if $JK[H_0]$ is nilpotent. Moreover H_0 is normal and has finite index in H so $JK[H]$ is nilpotent if and only if $JK[H_0]$ is nilpotent. The lemma is proved.

Lemma 16.9. Let H be a subgroup of G. Then $JK[G] \cap K[H] \subseteq JK[H]$.

Proof. Let $U = \{\alpha \in K[G]\} \mid \text{Supp } \alpha \cap H = \varnothing\}$. Then U is a left $K[H]$-module and $K[G] = K[H] + U$ is a direct sum. The result follows from Lemma 16.5.

The following are the basic results on the behavior of the Jacobson radical under algebraic field extensions.

Theorem 16.10. Let A be an algebra over a field K and let F be a field extension of K of finite degree $(F:K) = n$. Then

$$(J(F \otimes A))^n \subseteq F \otimes JA \subseteq J(F \otimes A).$$

If in addition F/K is separable, then $J(F \otimes A) = F \otimes JA$.

Proof. If $\{x_1 = 1, x_2, \ldots, x_n\}$ is a basis for F/K, then it is clearly also a normalizing basis for A^F over A with all $\sigma_i = 1$. Since $(JA) \cdot A^F$ is clearly equal to $F \otimes JA$, the result follows from Lemma 15.4 and Theorem 16.3.

Lemma 16.11. Let A be an algebra over a field K and let F be a field extension of K. Then $J(A^F) \cap A \subseteq JA$.

Proof. Write $F = K + F'$ where F' is a complementary K-subspace of F and set $U = F' \otimes A$. Then U is a left A-submodule of A^F and $A^F = A + U$, a direct sum. Thus by Lemma 16.5 we have $J(A^F) \cap A \subseteq JA$.

Lemma 16.12. Let A be an algebra over a field K and let F be a field extension of K. Let $\mathcal{L}$ be a family of fields such that

(i) $F \supseteq L \supseteq K$ for all $L \in \mathcal{L}$.

(ii) For any finite subset S of F, there exists $L \in \mathcal{L}$ with $S \subseteq L$.

Let $\alpha \in A$ and suppose that $\alpha \in J(A^L)$ for all $L \in \mathcal{L}$. Then $\alpha \in J(A^F)$.

Proof. Let $I = A^F \alpha A^F$ be the ideal of A^F generated by α and let $\beta \in I$. Then by (ii) above there exists $L \in \mathcal{L}$ with $\beta \in A^L \alpha A^L$. Now $\alpha \in J(A^L)$ and $J(A^L)$ is an ideal so $\beta \in J(A^L)$. Thus there exists $\gamma \in A^L \subseteq A^F$ with $\beta + \gamma + \beta\gamma = 0$. We have therefore shown that I is a quasi-regular ideal of A^F and thus $\alpha \in I \subseteq J(A^F)$.

Theorem 16.13 (Amitsur [*1*]). Let A be an algebra over a field K and let F be an algebraic field extension of K. Then $F \otimes JA \subseteq J(F \otimes A)$. If in addition F/K is separable, then $F \otimes JA = J(F \otimes A)$.

Proof. Let $\mathcal{L}$ be the family of all fields L with $F \supseteq L \supseteq K$ and $(L:K) < \infty$. Since F/K is algebraic, it follows that $\mathcal{L}$ satisfies (i) and (ii) of Lemma 16.12. Let $\alpha \in JA$. Then by Theorem 16.10, $\alpha \in J(A^L)$ for all $L \in \mathcal{L}$ and thus $\alpha \in J(A^F)$ by Lemma 16.12. This yields $J(A) \subseteq J(A^F)$ and hence $F \otimes J(A) \subseteq J(A^F)$ since the latter is an ideal.

Now suppose that F/K is separable and let $\beta \in J(A^F)$. Then for some $L \in \mathcal{L}$, $\beta \in A^L \cap J(A^F) \subseteq J(A^L)$ by Lemma 16.11 applied to the L-algebra A^L. Since L/K is a finite separable extension, we have $\beta \in J(A^L) = L \otimes JA$ by Theorem 16.10. Thus $\beta \in F \otimes JA$ so $J(A^F) \subseteq F \otimes JA$ and the result follows.

§17. ABELIAN EXTENSIONS

In this section, we continue our study of the relationship between JS and JR for certain rings S and subrings R. Here the proofs use fewer special properties of JS and many of the results hold for arbitrary characteristic ideals of S.

Let G be a group and let K be a field. Then we let $\mathrm{Hom}(G, K^*)$ denote the group of homomorphisms of G into the multiplicative group $K^* = K - \{0\}$. Multiplication in $\mathrm{Hom}(G, K^*)$ is defined pointwise so that if

λ, $\mu \in \mathrm{Hom}(G, K^*)$, then $(\lambda\mu)(x) = \lambda(x)\mu(x)$ for all $x \in G$. We say that G is K-complete if for each $x \in G$, $x \neq 1$, there exists $\lambda \in \mathrm{Hom}(G, K^*)$ with $\lambda(x) \neq 1$. We observe immediately that if G is K-complete then G is abelian and G has no elements of order p in case K has characteristic $p > 0$.

Lemma 17.1. Let G be an abelian group, let K be a field, and assume that G has no elements of order p in case K has characteristic $p > 0$.

(i) If K is algebraically closed, then G is K-complete.

(ii) Suppose $G = \prod G_\nu$ is a possibly infinite direct product of K-complete groups G_ν. Then G is K-complete.

(iii) If G is the direct product of infinite cyclic groups and if K is infinite, then G is K-complete.

(iv) If G is finitely generated, then there exists a separable algebraic extension F of K such that G is F-complete.

Proof. (i) Let $x \in G$, $x \neq 1$. Since G has no elements of order p if K has characteristic p and since K is algebraically closed, there exists $\lambda \in \mathrm{Hom}(\langle x \rangle, K^*)$ with $\lambda(x) \neq 1$. Since K is algebraically closed, K^* is divisible and hence injective. Thus λ extends to an element of $\mathrm{Hom}(G, K^*)$.

(ii) Let $x \in G = \prod G_\nu$ with $x \neq 1$. If $x = \prod x_\nu$ then for some ν, say $\nu = \nu_0$, we have $x_{\nu_0} \neq 1$. Now G_{ν_0} is K-complete so there exists $\lambda \in \mathrm{Hom}(G_{\nu_0}, K^*)$ with $\lambda(x_{\nu_0}) \neq 1$. We extend λ to an element of $\mathrm{Hom}(G, K^*)$ by composing it with the projection map $G \to G_{\nu_0}$.

(iii) By (ii) it suffices to show that the infinite cyclic group $\langle y \rangle$ is K-complete. Let $x \in \langle y \rangle$ with $x \neq 1$ so that $x = y^n$ for some $n \neq 0$. Now K is infinite so not every element a of K^* can satisfy $a^n = 1$. Let $a \in K^*$ with $a^n \neq 1$ and let $\lambda \in \mathrm{Hom}(\langle y \rangle, K^*)$ be given by $\lambda(y) = a$. Then $\lambda(x) = a^n \neq 1$.

(iv) If K is a finite field, let F be the algebraic closure of K. Then F/K is separable and G is F-complete by (i). Thus let K be infinite. Since G is finitely generated, the elements of finite order in G form a finite subgroup H and say $|H| = n$. Then $n \neq 0$ in K by assumption and we let $F = K(\epsilon)$ where ϵ is a primitive nth root of unity. Then F/K is separable. Now $G = \prod G_\nu$ is a finite direct product of cyclic groups and hence it suffices to show that each G_ν is F-complete. If G_ν is infinite, then this follows from (iii). Now suppose $G_\nu = \langle y \rangle$ is finite so that $y \in H$ and hence the order d of y divides n. Define $\lambda \in \mathrm{Hom}(\langle y \rangle, F^*)$ by $\lambda(y) = \epsilon^{n/d}$. Then λ is faithful on $\langle y \rangle$ so the result follows.

Lemma 17.2. Let G be K-complete and let $x_1, x_2, \ldots, x_n$ be distinct elements of G. Then there exists $\lambda_1, \lambda_2, \ldots, \lambda_n \in \mathrm{Hom}(G, K^*)$ such that the $n \times n$ matrix $(\lambda_i(x_j))$ is nonsingular.

Proof. Let $x_1, x_2, \ldots, x_n$ be distinct elements of G. We show first by induction on n that if $c_1, c_2, \ldots, c_n \in K$ and $c_1\lambda(x_1) + c_2\lambda(x_2) + \cdots + c_n\lambda(x_n) = 0$ for all $\lambda \in \mathrm{Hom}(G, K^*)$, then $c_1 = c_2 = \cdots = c_n = 0$. The result is clear for $n = 1$ since $\lambda(x_1) \neq 0$. Suppose the result is true for $n - 1$. If the above equation holds nontrivially, then clearly by induction all c_i are nonzero. Now $x_n \neq x_1$ so $x_n x_1^{-1} \neq 1$ and hence since G is K-complete there exists $\mu \in \mathrm{Hom}(G, K^*)$ with $\mu(x_n x_1^{-1}) \neq 1$ or $\mu(x_n) \neq \mu(x_1)$. Since $\mathrm{Hom}(G, K^*)$ is a group, it follows that $\sum c_i(\mu\lambda)(x_i) = 0$ for all $\lambda \in \mathrm{Hom}(G, K^*)$ so

$$c_1\mu(x_1)\lambda(x_1) + c_2\mu(x_2)\lambda(x_2) + \cdots + c_n\mu(x_n)\lambda(x_n) = 0.$$

Also multiplying the original equation by $\mu(x_n)$ yields

$$c_1\mu(x_n)\lambda(x_1) + c_2\mu(x_n)\lambda(x_2) + \cdots + c_n\mu(x_n)\lambda(x_n) = 0.$$

Subtracting one from the other, we have for all λ

$$\sum_1^{n-1} c_i\big(\mu(x_n) - \mu(x_i)\big)\lambda(x_i) = 0.$$

By induction $c_i\big(\mu(x_n) - \mu(x_i)\big) = 0$ and since $\mu(x_n) \neq \mu(x_1)$ we have $c_1 = 0$, a contradiction.

We now prove the lemma by induction on n. Again the result is clear for $n = 1$. By induction choose $\lambda_1, \lambda_2, \ldots, \lambda_{n-1}$ so that the $(n - 1) \times (n - 1)$ matrix $C_n = \big(\lambda_i(x_j)\big)$, $i, j = 1, 2, \ldots, n - 1$ is nonsingular. If we expand the $n \times n$ determinant $\det(\lambda_i(x_j))$ by cofactors with respect to the last row we have

$$\det\big(\lambda_i(x_j)\big) = c_1\lambda_n(x_1) + c_2\lambda_n(x_2) + \cdots + c_n\lambda_n(x_n)$$

where $c_n = \det C_n$ and hence $c_n \neq 0$ by the choice of $\lambda_1, \lambda_2, \ldots, \lambda_{n-1}$. Thus by the results of the preceding paragraph there exists $\lambda_n \in \mathrm{Hom}(G, K^*)$ with $\det\big(\lambda_i(x_j)\big) \neq 0$ and the result follows.

Let G be a group and let K be a field. If $\lambda \in \mathrm{Hom}(G, K^*)$, then λ defines a map $\lambda^*: K[G] \to K[G]$ given by

$$\lambda^*(\textstyle\sum a_x \cdot x) = \sum a_x\lambda(x) \cdot x.$$

It follows easily that λ^* is a K-automorphism of $K[G]$.

Lemma 17.3. Let G be a group, H a normal subgroup, and suppose that G/H is K-complete. Let $\alpha \in K[G]$ with

$$\alpha = \alpha_1 + \alpha_2 + \cdots + \alpha_n$$

where $\mathrm{Supp}\,\alpha_i \subseteq Hx_i$ and $x_1, x_2, \ldots, x_n$ are in distinct cosets of H.

Then there exists $\lambda_1, \lambda_2, \ldots, \lambda_n \in \mathrm{Hom}(G, K^*)$ and $k_1, k_2, \ldots, k_n \in K$ such that

$$\alpha_1 = k_1\lambda_1^*(\alpha) + k_2\lambda_2^*(\alpha) + \cdots + k_n\lambda_n^*(\alpha).$$

Proof. We consider $\mathrm{Hom}(G/H, K^*)$ as being naturally contained in $\mathrm{Hom}(G, K^*)$. Thus since $x_1, x_2, \ldots, x_n$ map into distinct elements of G/H and since G/H is K-complete, it follows from Lemma 17.2 that there exists $\lambda_1, \lambda_2, \ldots, \lambda_n \in \mathrm{Hom}(G/H, K^*)$ such that the $n \times n$ matrix $(\lambda_i(x_j))$ is nonsingular. Now $\mathrm{Supp}\, \alpha_j \subseteq Hx_j$ so $\lambda_i^*(\alpha_j) = \lambda_i(x_j)\alpha_j$. Thus

$$\lambda_i^*(\alpha) = \lambda_i(x_1)\alpha_1 + \lambda_i(x_2)\alpha_2 + \cdots + \lambda_i(x_n)\alpha_n.$$

Since the matrix $(\lambda_i(x_j))$ is nonsingular, the result follows.

Theorem 17.4. Let H be a normal subgroup of G and suppose that G/H is K-complete. Let I be an ideal in $K[G]$ which is invariant under all K-automorphisms of $K[G]$. Then

$$I = (I \cap K[H]) \cdot K[G].$$

Proof. Clearly $I \supseteq (I \cap K[H]) \cdot K[G]$. Now let $\alpha \in I$ and write $\alpha = \alpha_1 + \alpha_2 + \cdots + \alpha_n$ as in the previous lemma. Then by that lemma and the fact that $\lambda_i^*(I) \subseteq I$ we have $\alpha_i \in I$ and hence $\alpha_i x_i^{-1} \in I \cap K[H]$. Thus $\alpha \in (I \cap K[H]) \cdot K[G]$.

Corollary 17.5. Let G be a group and let K be a field. Then G is K-complete if and only if $K[G]$ has no ideals $I \neq 0$, $K[G]$ which are invariant under all K-automorphisms of $K[G]$.

Proof. Let G be K-complete and let I be an ideal of $K[G]$ invariant under all K-automorphisms of $K[G]$. Then by Theorem 17.4 with $H = \langle 1 \rangle$ we have $I = (I \cap K) \cdot K[G]$. Since $I \cap K = 0$ or K, we have $I = 0$ or $K[G]$.

Let $\mathcal{M}$ be the set of all ideals M of $K[G]$ such that $K[G]/M$ is K-isomorphic to K. Clearly any K-automorphism of $K[G]$ permutes the members of $\mathcal{M}$ and hence leaves invariant the ideal $I = \bigcap_{M \in \mathcal{M}} M$. Since the kernel of the natural map $K[G] \to K[G/G] \simeq K$ is in $\mathcal{M}$, we see that $I \neq K[G]$. Now suppose that G is not K-complete and choose $x \in G$, $x \neq 1$ such that $\lambda(x) = 1$ for all $\lambda \in \mathrm{Hom}(G, K^*)$. It then follows easily that $x - 1 \in I$ so $I \neq 0$.

Now Lemma 17.1 offers a large number of examples of groups which are K-complete. However it is certainly not true that every abelian group is K-complete. For example if $K = GF(2)$, then $|K^*| = 1$ and thus $\mathrm{Hom}(G, K^*) = \langle 1 \rangle$ for any group G. Therefore in this case G is K-complete if and only if $G = \langle 1 \rangle$.

Lemma 17.6. Let H be a normal subgroup of G and let $\mathscr{L}$ be a family of subgroups such that

(i) $G \supseteq L \supseteq H$ for all $L \in \mathscr{L}$.

(ii) For any finite subset S of G there exists $L \in \mathscr{L}$ with $S \subseteq L$.

Let $\alpha \in K[H]$ and suppose that $\alpha \in JK[L]$ for all $L \in \mathscr{L}$. Then $\alpha \in JK[G]$.

Proof. Let $I = K[G]\alpha K[G]$ be the ideal of $K[G]$ generated by α and let $\beta \in I$. Then by (ii) above there exists $L \in \mathscr{L}$ with $\beta \in K[L]\alpha K[L]$. Now $\alpha \in JK[L]$ and $JK[L]$ is an ideal, so $\beta \in JK[L]$. Thus there exists $\gamma \in K[L] \subseteq K[G]$ with $\beta + \gamma + \beta\gamma = 0$. We have therefore shown that I is a quasiregular ideal of $K[G]$ and thus $\alpha \in I \subseteq JK[G]$.

Theorem 17.7. Let H be a normal subgroup of G with G/H abelian and let K be a field. Suppose further that G/H has no elements of order p in case K has characteristic p. Then

$$JK[G] = (JK[G] \cap K[H]) \cdot K[G] \subseteq JK[H] \cdot K[G].$$

Proof. Clearly $JK[G] \supseteq (JK[G] \cap K[H]) \cdot K[G]$ and by Lemma 16.9, $JK[G] \cap K[H] \subseteq JK[H]$. It therefore suffices to show that $JK[G] \subseteq (JK[G] \cap K[H]) \cdot K[G]$. Fix $\alpha \in JK[G]$ and write $\alpha = \alpha_1 x_1 + \alpha_2 x_2 + \cdots + \alpha_n x_n$ where $\alpha_i \in K[H]$, $x_i \in G$, and $x_1, x_2, \ldots, x_n$ are in distinct cosets of H in G. We show that $\alpha_i \in JK[G]$ for all i. Let $\mathscr{L}$ be the family of all subgroups L of G such that $G \supseteq L \supseteq H$, L/H is finitely generated, and $x_1, x_2, \ldots, x_n \in L$. Clearly $\mathscr{L}$ satisfies (i) and (ii) of Lemma 17.6.

Let $L \in \mathscr{L}$. Then $\alpha \in JK[G] \cap K[L] \subseteq JK[L]$ by Lemma 16.9. Since L/H is a finitely generated abelian group with no elements of order p if K has characteristic p, it follows from Lemma 17.1(iv) that there exists a separable algebraic field extension F of K such that L/H is F-complete. Observe that $F \otimes_K K[L] = F[L]$ and hence by Theorem 16.13, $\alpha \in JF[L]$. Now $JF[L]$ is certainly invariant under all F-automorphisms of $F[L]$ and hence by Theorem 17.4 applied to $F[L]$ with $I = JF[L]$ we have $\alpha \in JF[L] = (JF[L] \cap F[H]) \cdot F[L]$. Thus $\alpha_i \in JF[L]$. Moreover since F/L is separable, Theorem 16.13 yields $JF[L] = F \otimes JK[L]$ so

$$\alpha_i \in K[L] \cap (F \otimes JK[L]) = JK[L].$$

Therefore Lemma 17.6 applies to each α_i and we conclude that $\alpha_i \in JK[G] \cap K[H]$ and hence $\alpha \in (JK[G] \cap K[H]) \cdot K[G]$. This completes the proof.

Applying the above with $H = \langle 1 \rangle$ yields

Corollary 17.8. Let G be an abelian group with no elements of order p in case K has characteristic p. Then $JK[G] = 0$.

We now observe that similar results hold in the case of purely transcendental field extensions. Let F be a purely transcendental extension of K with transcendence basis $\{x_\nu\}$ and let $G = \langle x_\nu \mid \text{all } \nu\rangle$ be the multiplicative subgroup of F^* generated by this basis. If $K \cdot G \subseteq F$ denotes the set of all finite sums $\sum a_x \cdot x$ with $a_x \in K$, $x \in G$, then since $\{x_\nu\}$ is algebraically independent over K we have $K \cdot G \simeq K[G]$ and $G = \prod_\nu \langle x_\nu\rangle$ is a direct product of infinite cyclic groups. Let $\lambda \in \operatorname{Hom}(G, K^*)$ so that λ induces an automorphism λ^* on $K[G]$. Since $K[G] \simeq K \cdot G$, λ^* can be viewed as an automorphism of $K \cdot G$ and then finally as one of F, since F is clearly the quotient field of $K \cdot G$. Thus λ^* is a field automorphism of F which fixes K.

Theorem 17.9. Let A be an algebra over an infinite field K and let F be a purely transcendental field extension of K. If I is an ideal of $F \otimes A$ which is invariant under all ring automorphisms of $F \otimes A$, then $I = F \otimes (I \cap A)$.

Proof. Clearly $I \supseteq F \otimes (I \cap A)$. Now let $\alpha \in I$ and use the above notation. Then there clearly exists $a \in F$, $a \neq 0$ with

$$a\alpha = y_1\alpha_1 + y_2\alpha_2 + \cdots + y_n\alpha_n$$

where $\alpha_i \in A$ and $y_1, y_2, \ldots, y_n$ are distinct elements of G. By Lemmas 17.1(iii) and 17.2 there exists $\lambda_1, \lambda_2, \ldots, \lambda_n \in \operatorname{Hom}(G, K^*)$ with the $n \times n$ matrix $(\lambda_i(y_j))$ nonsingular. As above each λ_i induces a field automorphism λ_i^* of F fixing K and hence λ_i^* extends in a natural way to a ring automorphism of $F \otimes A$. We have

$$\lambda_i^*(a\alpha) = \lambda_i(y_1)y_1\alpha_1 + \lambda_i(y_2)y_2\alpha_2 + \cdots + \lambda_i(y_n)y_n\alpha_n$$

and $\lambda_i^*(a\alpha) \in I$ since I is an ideal invariant under all such automorphisms. Moreover since $(\lambda_i(y_j))$ is nonsingular, we can solve for the $y_i\alpha_i$ and we conclude that $y_i\alpha_i \in I$. Thus $a^{-1}\alpha_i \in I \cap A$ and $\alpha \in F \otimes (I \cap A)$.

Theorem 17.10 (Amitsur [*1*]). Let A be an algebra over a field K and let F be a purely transcendental field extension of K. Then

$$J(F \otimes_K A) = F \otimes_K \big(J(F \otimes_K A) \cap A\big).$$

If in addition $F \neq K$, then $J(F \otimes_K A) \cap A$ is a nil ideal of A.

Proof. We show first that $J(F \otimes A) = F \otimes \big(J(F \otimes A) \cap A\big)$. This of course follows immediately from Theorem 17.9 in case K is infinite. Thus let us assume that K is finite. Let L be the algebraic closure of F, let $\tilde{K}$ be the algebraic closure of K in L, and let $\tilde{F}$ be the composite field $\tilde{F} = F \cdot \tilde{K}$. Note that $\tilde{K}/K$ is separable since K is finite. Moreover since

$\tilde{K}$ and F are linearly disjoint over K, it follows that $\tilde{F}/F$ is separable and algebraic and $\tilde{F}/\tilde{K}$ is purely transcendental. Of course $\tilde{K}$ is infinite. Set $B = F \otimes A = A^F$.

Let $\alpha \in J(F \otimes A)$ and write $\alpha = a_1\alpha_1 + a_2\alpha_2 + \cdots + a_n\alpha_n$ where $\alpha_i \in A$ and $a_1, a_2, \ldots, a_n$ are elements of F which are linearly independent over K. By Theorem 16.13, $\alpha \in J(\tilde{F} \otimes_F B) = J(\tilde{F} \otimes_{\tilde{K}} A^{\tilde{K}})$. Since $\tilde{F}/\tilde{K}$ is purely transcendental and $\tilde{K}$ is infinite and since $a_1, a_2, \ldots, a_n$ are elements of $\tilde{F}$ linearly independent over $\tilde{K}$, we have by the above $\alpha_i \in J(\tilde{F} \otimes B)$. Thus since $\tilde{F}/F$ is separable, Theorem 16.13 yields $\alpha_i \in \big(\tilde{F} \otimes_F J(B)\big) \cap B = J(B)$. Thus $\alpha_i \in J(F \otimes_K A) \cap A$ and $\alpha \in F \otimes_K \big(J(F \otimes A) \cap A\big)$. We have therefore shown that $J(F \otimes A) \subseteq F \otimes \big(J(F \otimes A) \cap A\big)$ and since the reverse inclusion is trivially true, this first result follows.

Suppose now that $F \neq K$. We show that $J(F \otimes A) \cap A$ is a nil ideal of A. Fix $x \in F$ with x transcendental over K and let $\alpha \in J(F \otimes A) \cap A$. If $\alpha = 0$, then certainly α is nilpotent so we may assume that $\alpha \neq 0$. Now by Lemma 16.11 applied to the $K(x)$-algebra $\bar{A} = K(x) \otimes A$ we have $\alpha x \in J(F \otimes \bar{A}) \cap \bar{A} \subseteq J(\bar{A})$ and thus there exists $\gamma \in \bar{A}$ with $\alpha x + \gamma + \alpha x \gamma = 0$. Since $K(x)$ is the quotient field of the polynomial ring $K[x]$, we have clearly $\gamma = a^{-1}\beta$ where $a = \sum_{i=0} a_i x^i \in K[x]$, $\beta = \sum_{i=0} \beta_i x^i$ with $a_i \in K$, $\beta_i \in A$. Of course these sums are finite and $a \neq 0$. Now $\alpha x + \gamma + \alpha x \gamma = 0$ yields $\alpha x a + \beta + \alpha x \beta = 0$ and hence by considering the coefficient of x^i here we have $\beta_0 = 0$ and $\beta_i = -\alpha(\beta_{i-1} + a_{i-1})$ for $i > 0$. By assumption $\alpha x a \neq 0$ so $\beta \neq 0$; and thus let j be minimal with $\beta_j \neq 0$. It follows from the above that $j > 0$ and that $\beta_j = -a_{j-1}\alpha$ and hence by induction for $i \geq j$ that β_i is a polynomial over K in α of degree precisely $i - j + 1$. Since $\sum_{i=0} \beta_i x^i$ is a finite sum there exists $i > j$ with $\beta_i = 0$ and therefore α is algebraic over K. Now we can clearly assume that α satisfies a polynomial equation of the form $\alpha^n(1 + c_1\alpha + c_2\alpha^2 + \cdots) = 0$. Since $\alpha \in J(F \otimes A)$ we have $c_1\alpha + c_2\alpha^2 + \cdots \in J(F \otimes A)$ and hence $1 + c_1\alpha + c_2\alpha^2 + \cdots$ has a right inverse in $F \otimes A$. This implies that $\alpha^n = 0$ and the result follows.

§18. SEMISIMPLE RINGS

In this section we offer certain fairly general conditions which imply that the group algebra $K[G]$ is semisimple.

Let A be an algebra over a field K. We say that A is nilpotent free if for all fields $F \supseteq K$, $F \otimes A = A^F$ has no nonzero nilpotent ideals.

Lemma 18.1. Let A be a nilpotent free algebra over K and let F and L be fields with $F \supseteq L \supseteq K$.

(i) If $J(A^L) = 0$, then $J(A^F) = 0$.

(ii) If F/L is algebraic, then $J(A^L) = 0$ if and only if $J(A^F) = 0$.

Proof. (i) Let F_0 be a subfield of F with $F \supseteq F_0 \supseteq L$, F_0/L purely transcendental, and F/F_0 algebraic. By Lemma 16.11 and Theorem 17.10 applied to the L-algebra A^L, we have $J(A^{F_0}) = 0$. Let $\alpha \in J(A^F)$. Since F/F_0 is algebraic, it follows that there exists a field F_1 with $F \supseteq F_1 \supseteq F_0$, $(F_1 : F_0) = n < \infty$, and $\alpha \in A^{F_1}$. Hence $\alpha \in J(A^F) \cap A^{F_1} \subseteq J(A^{F_1})$ by Lemma 16.11. By Theorem 16.10 we have $J(A^{F_1})^n \subseteq F_1 \otimes J(A^{F_0}) = 0$ and thus since A^{F_1} has no nonzero nilpotent ideals we have $J(A^{F_1}) = 0$. This yields $\alpha = 0$ and (i) follows.

(ii) By (i) above $J(A^L) = 0$ imples $J(A^F) = 0$. On the other hand if $J(A^L) \neq 0$ then by Theorem 16.13 applied to the L-algebra A^L we have $F \otimes J(A^L) \subseteq J(A^F)$ so $J(A^F) \neq 0$.

Let t.d. (F/K) denote the transcendence degree of F/K so that t.d. (F/K) is either a nonnegative integer or the symbol ∞.

Theorem 18.2. Let A be a nilpotent free algebra over a field K and let F and F' be two field extensions of K with t.d. $(F'/K) \geq$ t.d. (F/K). Then $J(A^F) = 0$ implies that $J(A^{F'}) = 0$.

Proof. By Lemma 18.1(ii) it suffices to assume that F/K and F'/K are purely transcendental. By Theorem 17.10 it suffices to show that $J(A^{F'}) \cap A = 0$. Let $\alpha \in J(A^{F'}) \cap A$ and let $\mathscr{L}$ be the set of all fields L with $F \supseteq L \supseteq K$ and L/K purely transcendental with finite transcendence degree. Since F/K is purely transcendental, if follows that $\mathscr{L}$ satisfies (i) and (ii) of Lemma 16.12. Let $L \in \mathscr{L}$. Since t.d. $(F'/K) \geq$ t.d. (F/K), it follows that there exists a field L' with $F' \supseteq L' \supseteq K$ and such that L and L' are K-isomorphic. Thus A^L and $A^{L'}$ are A-isomorphic. Now $\alpha \in J(A^{F'}) \cap A^{L'} \subseteq J(A^{L'})$ by Lemma 16.11 so it follows that $\alpha \in J(A^L)$. Since this is true for all such $L \in \mathscr{L}$, Lemma 16.12 yields $\alpha \in J(A^F)$ and hence $\alpha = 0$. This completes the proof.

The semisimplicity of the complex group algebra $C[G]$ was first proved by Rickart [*48*] by analytic means. Later Amitsur [*1*] and Herstein (unpublished) showed independently that if K is a nondenumerable field of characteristic 0, then $K[G]$ is semisimple. The sharpest result to date is Theorem 18.3 below (also due to Amitsur). We will prove the following two theorems simultaneously.

Theorem 18.3 (Amitsur [*2*]). Let K be a field of characteristic 0 which is not algebraic over the rational field Q. If G is a group, then $K[G]$ is semisimple.

Theorem 18.4 (Passman [*41*]). Let K be a field of characteristic $p > 0$ which is not algebraic over the prime field $GF(p)$. If G is a group with no elements of order p, then $K[G]$ is semisimple.

Proof. Let K_0 be the prime subfield of K. We first observe that if F is any field containing K_0 then $F \otimes_{K_0} K_0[G] = F[G]$. Thus by Lemma 3.1 and Theorems 3.3 and 3.7, $K_0[G]$ is nilpotent free. Let F be a purely transcendental field extension of K_0 with t.d. $(F/K_0) = 1$. By Theorem 17.10, $JF[G] = F \otimes I$ where I is a nil ideal in $K_0[G]$. Thus by Theorems 3.2 and 3.6 we have $I = 0$ and $JF[G] = 0$. Now by assumption t.d. $(K/K_0) \geq$ t.d. (F/K_0) so the result follows from Theorem 18.2.

The question of whether the above results hold without the assumption that t.d. $(K/K_0) \geq 1$ is probably the most interesting and most difficult of the unanswered problems in the field. We can now completely answer the question of the existence of nil ideals in the characteristic 0 case.

Theorem 18.5. Let K be a field of characteristic 0 and let G be a group. Then $K[G]$ contains no nonzero nil ideals.

Proof. Let Q denote the field of rational numbers. If t.d. $(K/Q) \geq 1$, then by Theorem 18.3 $JK[G] = 0$ and hence certainly $K[G]$ has no nil ideals. Thus we may assume that K is algebraic over Q. Suppose by way of contradiction that I is a nonzero nil ideal in $K[G]$ and let $\alpha \in I$ with $\alpha \neq 0$. If $x \in \operatorname{Supp} \alpha$, then since I is an ideal $x^{-1}\alpha \in I$ and $1 \in \operatorname{Supp} x^{-1}\alpha$. Thus by replacing α by $x^{-1}\alpha$ if necessary we may assume that $1 \in \operatorname{Supp} \alpha$.

Let $\alpha = \sum a_x \cdot x$ and let F be the field extension of Q generated by the finite set $\{a_x \mid x \in \operatorname{Supp} \alpha\}$. Since K/Q is algebraic, it follows that F/Q is finite; let A denote the ring of algebraic integers in F. Then multiplying α by a suitable nonzero element of F if necessary we may assume that $a_x \in A$ for all x. Now $a_1 \neq 0$ and hence there exists only finitely many ideals of A which contain a_1. Thus there exists a rational integral prime p such that a_1 is in no proper ideal of A which contains p and such that p is larger than the orders of the elements of finite order in $\operatorname{Supp} \alpha$. Let M be a maximal ideal of A containing p.

Now the map $A \to \bar{A} = A/M$ clearly induces a homomorphism of $A \cdot G$ onto $\bar{A}[G]$ where $A \cdot G = \{\sum b_x x \in K[G] \mid b_x \in A\}$. Since $\alpha \in A \cdot G$, it follows that its image $\bar{\alpha} = \sum \bar{a}_x x \in \bar{A}[G]$ is nilpotent. By the choice of M we see that $\bar{A}$ is a field of characteristic p. Moreover since $a_1 \notin M$, we have $\bar{a}_1 \neq 0$ and $1 \in \operatorname{Supp} \bar{\alpha}$. Thus by Lemma 3.5, there exists $x \in \operatorname{Supp} \bar{\alpha} \subseteq \operatorname{Supp} \alpha$, $x \neq 1$ such that the order of x is a power of p. This is a contradiction since p was chosen to be larger than all such finite orders. Hence $I = 0$ and $K[G]$ has no nonzero nil ideals.

Since Theorems 18.3 and 18.4 require t.d. $(K/K_0) \geq 1$, where K_0 is the prime subfield of K, we now consider a number of special cases in which the result is field independent.

Lemma 18.6. Let K be a field, G a group, and let $\mathscr{F}$ be a family of subgroups of G. Suppose that

(i) If $H \in \mathscr{F}$, then $K[H]$ is semisimple.

(ii) If S is a finite subset of G, then there exists $H \in \mathscr{F}$ with $S \subseteq H$.

Then $K[G]$ is semisimple.

Proof. Let $\alpha \in JK[G]$. Since $\operatorname{Supp} \alpha$ is a finite set, it follows from (ii) that there exists $H \in \mathscr{F}$ with $\alpha \in K[H]$. Thus $\alpha \in JK[G] \cap K[H] \subseteq JK[H] = 0$ by (i) above and by Lemma 16.9. The result follows.

Theorem 18.7. Let G be a locally finite group and let K be a field. Suppose further that G has no elements of order p in case K has characteristic p. Then $K[G]$ is semisimple.

Proof. Let $\mathscr{F}$ be the family of all finite subgroups of G. Then by assumptions and Theorem 15.3 the hypotheses of Lemma 18.6 are satisfied and this yields the result.

Lemma 18.8. Let N be a normal subgroup of G with $G/N = B$ abelian. Suppose that $K[N]$ is semisimple and that G has no elements of order p in case K has characteristic p. Then $K[G]$ is semisimple.

Proof. We suppose first that B is finitely generated so that B_1, the torsion subgroup of B, is finite of order n for some integer n. Let $G \supseteq G_1 \supseteq N$ with $G_1/N = B_1$. Then Theorem 16.6 yields $(JK[G_1])^n \subseteq (JK[N]) \cdot K[G_1] = 0$ and hence $JK[G_1]$ is nilpotent. Thus by Lemma 3.1 and Theorems 3.3 and 3.6 we have $JK[G_1] = 0$. Now G_1 is normal in G and G/G_1 is a torsion free abelian group so $JK[G] \subseteq (JK[G_1]) \cdot K[G] = 0$ by Theorem 17.7 and the result follows in this case.

Now let $\mathscr{F}$ be the family of all subgroups H of G such that $G \supseteq H \supseteq N$ and H/N is finitely generated. It follows from the above that the hypotheses of Lemma 18.6 are satisfied and hence $JK[G] = 0$ in general.

Theorem 18.9 (Villamayor [*54*], Wallace [*60*], Passman [*41*], Zalessky [*62*]). Let G be a solvable group and let K be a field. Suppose further that

G has no elements of order p in case K has characteristic p. Then $K[G]$ is semisimple.

Proof. Let $G = G_n \supseteq G_{n-1} \supseteq \cdots \supseteq G_1 \supseteq G_0 = \langle 1 \rangle$ be a subinvariant series for G with G_i/G_{i-1} abelian. We prove by induction on $i = 0, 1, \ldots, n$ that $JK[G_i] = 0$. Since $G_0 = \langle 1 \rangle$, the case $i = 0$ is clear. Suppose now that $i \leq n$ is given and that we have already shown that $JK[G_{i-1}] = 0$. Since G_i/G_{i-1} is abelian and G_i has no elements of order p in case K has characteristic p, it follows from Lemma 18.8 that $JK[G_i] = 0$. In particular with $i = n$ we have $JK[G] = 0$.

Theorem 18.10 (Amitsur [*2*]). Let $K[G]$ satisfy a polynomial identity and suppose further that $\Delta(G)$ has no elements of order p in case K has characteristic p. Then $K[G]$ is semisimple.

Proof. Let $H = \mathfrak{S}(\Delta(G))$ so that H is a locally solvable subgroup of G having no elements of order p in case K has characteristic p. Thus by Lemma 18.6 and Theorem 18.9 we have $JK[H] = 0$. Now Theorems 5.5 and 13.6 imply that $[G:H] < \infty$ and hence by Theorem 16.6, $JK[G]$ is nilpotent. Thus by Lemma 3.1 and Theorems 3.3 and 3.6 we have $JK[G] = 0$.

§19. LOCALLY FINITE GROUPS

In this and the next section we will study the nilpotent ideals in group rings. Since there are no nonzero nilpotent ideals in characteristic 0, we will assume throughout this section that all fields have characteristic $p > 0$. We start with two lemmas on finite groups.

Lemma 19.1. Let K be an algebraically closed field, let G be a finite group, and let Z be a central subgroup of G. Let I be an ideal in $K[Z]$ with $K[Z]/I = K$ and suppose that the algebra $A = K[G]/I \cdot K[G]$ is semisimple. Then $p \nmid [G:Z]$.

Proof. We remark first that since Z is central in G, $I \cdot K[G]$ is an ideal of $K[G]$ so that A does in fact exist. If $\{x_1, x_2, \ldots, x_n\}$ is a complete set of coset representatives for Z in G, then it follows easily that $I \cdot K[G] = Ix_1 + Ix_2 + \cdots + Ix_n$, a vector space direct sum. Since $K[Z]/I = K$, we then have $\dim_K A = [G:Z]$.

Let H be a subgroup of G such that $G \supseteq H \supseteq Z$ and H/Z is a Sylow p-subgroup of G/Z and say $|H/Z| = p^a$. From the above we have easily $I \cdot K[G] \cap K[H] = I \cdot K[H]$ and thus if B is the image of $K[H]$ in A, then $B = K[H]/I \cdot K[H]$ and thus $\dim B = [H:Z] = p^a$. Let $\{y_1, y_2, \ldots, y_t\}$ be a complete set of left coset representatives for H in G with

$t = [G:H]$. Since $K[G] = \sum y_i K[H]$, it follows that $A = \sum \bar{y}_i B$ where $\bar{y}_i$ is the image of y_i in A. By dimension considerations, the latter must be a direct sum and hence A is a finitely generated free right B-module.

If P is a Sylow p-subgroup of H, then $H = ZP$. Since $K[Z]/I = K$ it therefore follows that B is a homomorphic image of $K[P]$. By Lemma 10.1(ii), $JK[P]$ has codimension 1 in $K[P]$ and hence it is its unique maximal right ideal. This implies that B has a unique maximal right ideal and thus B is an indecomposable right B-module.

Let M be an irreducible A-module of dimension f. Then since A is a finite dimensional semisimple algebra we have $A \simeq M + M'$, a direct sum of A-modules. Since A is a finitely generated free B-module, we then have as B-modules

$$M_B + M'_B \simeq A_B \simeq \sum B_B$$

where the latter is a finite direct sum of copies of B. Now B is an indecomposable B-module and thus by the Krull–Schmidt theorem we have $M_B = \sum' B_B$ as B-modules and thus $\dim B = p^a$ divides f.

Now K is algebraically closed and A is semisimple so we have

$$[G:Z] = \dim A = \sum f_i^2$$

where each f_i is the dimension of an irreducible A-module. By the above $p^a \mid f_i$ so $p^{2a} \mid f_i^2$ and hence $p^{2a} \mid [G:Z]$. Thus $p^{2a} \mid p^a$ since H/Z is a Sylow p-subgroup of G/Z and therefore we have $p^a = 1$.

Lemma 19.2. Let G be a finite group and let Z be a central subgroup. Suppose that there exists $\alpha \in K[Z]$, $\alpha \neq 0$ with $\alpha JK[G] = 0$. Then $p \nmid [G:Z]$.

Proof. By multiplying α by a field element if necessary we may assume that some coefficient in α is equal to 1. Let $K_0 = GF(p)$ be the prime subfield of K. Since G is finite, $JK_0[G]$ is nilpotent and hence $K \cdot JK_0[G]$ is a nilpotent ideal in $K[G]$. Thus $JK_0[G] \subseteq JK[G]$ and hence $\alpha JK_0[G] = 0$. Write $K = K_0 + K_1$, a direct sum of K_0-vector spaces and then write $\alpha = \alpha_0 + \alpha_1$ where $\alpha_0 \in K_0[G]$ and all coefficients of α_1 are contained in in K_1. By assumption, $\alpha_0 \in K_0[Z]$ and $\alpha_0 \neq 0$. It now follows easily that $\alpha_0 JK_0[G] = 0$.

Let $\tilde{K}_0$ be the algebraic closure of K_0. Since K_0 is perfect, $\tilde{K}_0/K_0$ is separably algebraic and hence by Theorem 16.3, $J\tilde{K}_0[G] = \tilde{K}_0 \cdot JK_0[G]$. Thus $\alpha_0 J\tilde{K}_0[G] = 0$. We have therefore shown that there exists $\alpha_0 \in \tilde{K}_0[Z]$, $\alpha_0 \neq 0$ with $\alpha_0 J\tilde{K}_0[G] = 0$. Thus by replacing K by $\tilde{K}_0$ if necessary we may assume that K is algebraically closed.

Let $B = \{\beta \in K[Z] \mid \alpha\beta = 0\}$ so since Z is abelian and $\alpha \neq 0$ it follows that B is an ideal in $K[Z]$ with $B \neq K[Z]$. Let I be a maximal ideal of $K[Z]$ with $I \supseteq B$. Then $K[Z]/I$ is a field and hence $K[Z]/I = K$ since K is algebraically closed. Let $\gamma \in JK[G]$ and write $\gamma = \sum \beta_i x_i$ where $\beta_i \in K[Z]$ and $\{x_i\}$ is a set of coset representatives for Z in G. By assumption

$$0 = \alpha\gamma = \sum (\alpha\beta_i)x_i$$

and hence clearly $\alpha\beta_i = 0$ for all i. Thus $\beta_i \in B \subseteq I$ and $JK[G] \subseteq I \cdot K[G]$. It then follows that $A = K[G]/I \cdot K[G]$ is a homomorphic image of the finite dimensional semisimple algebra $K[G]/JK[G]$ and thus by the Wedderburn theorems A is also semisimple. Then Lemma 19.1 yields $p \nmid [G:Z]$ and the result follows.

Let G be an arbitrary group. We define

$\Delta^+ = \Delta^+(G) = \{x \in \Delta(G)\} \mid x \text{ has finite order}\}$.

$\Delta^p = \Delta^p(G) = \langle x \in \Delta(G) \mid x \text{ has order a power of } p\rangle$.

Lemma 19.3. Let G be a group.

(i) Δ^+ is a normal subgroup of G and Δ/Δ^+ is torsion free abelian.

(ii) If S is any finite subset of Δ^+, then there exists a finite normal subgroup H of G with $S \subseteq H$.

(iii) Δ^p is a normal subgroup of G and Δ^+/Δ^p is a locally finite group having no elements of order p.

(iv) If H is a finite normal subgroup of G, then $\Delta^+(G/H) = \Delta^+(G)/H$ and $\Delta^p(G/H) = \Delta^p(G)H/H$.

(v) If $W \subseteq \Delta^p(G)$ with $[\Delta^p(G): W] < \infty$, then $[W:\Delta^p(W)] < \infty$.

Proof. Let S be a finite subset of $\Delta^+ \subseteq \Delta$ and let H be the subgroup of G generated by the elements of S and their finitely many conjugates in G. Then H is a normal subgroup of G, $S \subseteq H$, and H is a finitely generated subgroup of Δ. By Lemma 2.2, H' is finite. Now H/H' is a finitely generated abelian group, generated by elements of finite order. Thus H/H' is finite and so H is finite and (ii) follows. This also clearly implies that Δ^+ is a subgroup of G which is obviously normal.

Let $x, y \in \Delta$ and consider the group $\langle x, y\rangle$. Then $\langle x, y\rangle$ is a finitely generated subgroup of Δ so by Lemma 2.2, $\langle x, y\rangle'$ is finite. Hence $(x, y) = x^{-1}y^{-1}xy \in \Delta^+$. Since this is true for all $x, y \in \Delta$ we see that Δ/Δ^+ is abelian. Moreover since Δ^+ consists of all elements of Δ of finite order, we see that Δ/Δ^+ is torsion free and (i) follows.

In view of (ii) above, statements (iii) and (iv) are obvious. We now consider (v). Since $[\Delta^p(G): W] < \infty$, it follows from (ii) that there exists a finite normal subgroup H of G such that $\Delta^p(G) = WH$. By (iv)

$$\Delta^p(G/H) = WH/H \simeq W/(W \cap H) = \overline{W}$$

and so $\Delta^p(\overline{W}) = \overline{W}$. By (iv) again

$$\overline{W} = \Delta^p(\overline{W}) = (\Delta^p(W) \cdot (W \cap H))/(W \cap H)$$

and so $W = \Delta^p(W) \cdot (W \cap H)$. Since $|W \cap H| < \infty$, the result follows.

If R is a ring, we let NR denote its nilpotent radical. Thus NR is the sum of all the nilpotent ideals of R. We now consider the radicals associated with the groups $\Delta(G)$, $\Delta^+(G)$, and $\Delta^p(G)$.

Lemma 19.4. Let H be a normal subgroup of G with $H \subseteq \Delta^+(G)$. Then

$$(JK[H]) \cdot K[G] \subseteq NK[G]$$

and

$$JK[H] = NK[H] = \cup\, JK[W]$$

where the union is taken over all finite normal subgroups W of G contained in H.

Proof. Let W be a finite normal subgroup of G with $W \subseteq H$. Then $JK[W]$ is nilpotent so by Lemma 16.7 $(JK[W]) \cdot K[G]$ and $(JK[W]) \cdot K[H]$ are nilpotent ideals in $K[G]$ and $K[H]$, respectively. This implies that $\cup\, JK[W] \subseteq NK[H] \subseteq JK[H]$ and $(JK[W]) \cdot K[G] \subseteq NK[G]$. On the other hand, let $\alpha \in JK[H]$. Since Supp α is a finite subset of H, there exists by Lemma 19.3(ii) a finite normal subgroup W of G with $W \subseteq H$ and $\alpha \in K[W]$. Thus $\alpha \in JK[H] \cap K[W] \subseteq JK[W]$ by Lemma 16.9. Hence $JK[H] \subseteq \cup\, JK[W]$ and the result follows.

Lemma 19.5. Let $G = \Delta^p(G)$ and suppose that there exists $\alpha \in K[G]$, $\alpha \neq 0$ with $\alpha JK[G] = 0$. Then G is finite.

Proof. Since Supp α is finite, there exists by Lemma 19.3(ii) a finite normal subgroup H of G with $\alpha \in K[H]$. Let $C = \mathbf{C}_G(H)$ and put $W = \Delta^p(C)$. By Lemma 19.3(v), $[G: W] < \infty$. Also W is normal in G and $W = \Delta^p(W)$. Set $Z = H \cap W$ so that Z is a finite central subgroup of W. We show now that $W = Z$.

Let $x \in W$ have order a power of p and by Lemma 19.3(ii) let $\bar{G}$ be a finite normal subgroup of G with $Z \subseteq \bar{G} \subseteq W$ and $x \in \bar{G}$. By Lemma 19.4 we have $JK[\bar{G}] \subseteq JK[G]$ and hence $\alpha JK[\bar{G}] = 0$. Since $\alpha \in K[H]$, we can write $\alpha = \sum x_i \alpha_i$ where $\alpha_i \in K[Z]$ and the x_i are in distinct cosets of Z

in H. Hence the x_i are in distinct cosets of $\bar{G}$ in G since $\bar{G} \cap H = Z$. Thus $0 = \alpha JK[\bar{G}] = \sum x_i \alpha_i JK[\bar{G}]$ implies that $\alpha_i JK[\bar{G}] = 0$. Now $\alpha_i \neq 0$ for some i and Z is a central subgroup of $\bar{G}$ so Lemma 19.2 implies that $p \nmid [\bar{G}:Z]$. This yields $x \in Z$ and since $W = \Delta^p(W)$ is generated by its elements of order a power of p we have $W = Z$. Thus W is finite and since $[G:W]$ is finite we see that G is finite and the result follows.

Lemma 19.6. Let $G = \Delta(G)$. Then

$$JK[G] = NK[G] = (NK[\Delta^p(G)]) \cdot K[G].$$

Proof. By Lemma 19.4 we have

$$(NK[\Delta^p]) \cdot K[G] \subseteq NK[G] \subseteq JK[G]$$

and $NK[\Delta^p] = JK[\Delta^p]$. Thus it suffices to show that $JK[G] \subseteq (JK[\Delta^p]) \cdot K[G]$.

Now by Lemma 19.3(i), $G/\Delta^+(G)$ is torsion free abelian and hence by Theorem 17.7, $JK[G] \subseteq (JK[\Delta^+]) \cdot K[G]$. Let $\alpha \in JK[\Delta^+]$. Then by Lemma 19.3(iii), there exists a subgroup H of Δ^+ such that $H \supseteq \Delta^p$, $[H:\Delta^p] < \infty$, and $\alpha \in K[H]$. Thus $\alpha \in JK[\Delta^+] \cap K[H] \subseteq JK[H]$ by Lemma 16.9. Now by Lemma 19.3(iii), $p \nmid [H:\Delta^p]$ and therefore Theorem 16.6 yields

$$\alpha \in (JK[\Delta^p]) \cdot K[H] \subseteq (JK[\Delta^p]) \cdot K[\Delta^+].$$

Since $\alpha \in JK[\Delta^+]$ was arbitrary, we have $JK[\Delta^+] \subseteq (JK[\Delta^p]) \cdot K[\Delta^+]$ so

$$JK[G] \subseteq (JK[\Delta^+]) \cdot K[G] \subseteq (JK[\Delta^p]) \cdot K[G]$$

and the lemma is proved.

§20. THE NILPOTENT RADICAL

We can now quickly obtain our characterization of $NK[G]$, the nilpotent radical of $K[G]$.

Lemma 20.1. Let θ denote the natural projection $\theta: K[G] \to K[\Delta]$.

(i) Let $\alpha_1, \alpha_2, \ldots, \alpha_n \in K[G]$ and suppose that $\alpha_1 x_1 \alpha_2 x_2 \cdots \alpha_{n-1} x_{n-1} \alpha_n = 0$ for all $x_1, x_2, \ldots, x_{n-1} \in G$. Then $\theta(\alpha_1)\theta(\alpha_2) \cdots \theta(\alpha_n) = 0$.

(ii) Let $A_1, A_2, \ldots, A_n$ be ideals of $K[G]$ and suppose that $A_1 A_2 \cdots A_n = 0$. Then the sets $\theta(A_i)$ are ideals in $K[\Delta]$ and $\theta(A_1)\theta(A_2) \cdots \theta(A_n) = 0$.

Proof. (i) We show by induction on n that if $\alpha_1 x_1 \alpha_2 x_2 \cdots \alpha_{n-1} x_{n-1} \alpha_n = 0$ for all $x_i \in G$, then $\theta(\alpha_1)\theta(\alpha_2) \cdots \theta(\alpha_{n-1})\alpha_n = 0$. The case $n = 1$ is trivial. Thus let $n > 1$ and suppose the result is true for $n - 1$. Fix $x_{n-1} \in G$ and set $\beta = \alpha_{n-1} x_{n-1} \alpha_n$. Then for all $x_1, x_2, \ldots, x_{n-2} \in G$ we have $\alpha_1 x_1 \alpha_2 x_2 \cdots \alpha_{n-2} x_{n-2} \beta = 0$ and thus by induction

$$\theta(\alpha_1)\theta(\alpha_2) \cdots \theta(\alpha_{n-2})\alpha_{n-1} x_{n-1} \alpha_n = \theta(\alpha_1)\theta(\alpha_2) \cdots \theta(\alpha_{n-2})\beta = 0.$$

Now the above holds for all $x_{n-1} \in G$ so by Lemma 1.3 with $\gamma = \delta = 0$ we have

$$\theta(\alpha_1)\theta(\alpha_2) \cdots \theta(\alpha_{n-1})\alpha_n = \theta\big(\theta(\alpha_1)\theta(\alpha_2) \cdots \theta(\alpha_{n-2})\alpha_{n-1}\big)\alpha_n = 0$$

and the inductive result is proved. Finally $\theta(\alpha_1)\theta(\alpha_2) \cdots \theta(\alpha_{n-1})\alpha_n = 0$ implies easily that $\theta(\alpha_1)\theta(\alpha_2) \cdots \theta(\alpha_n) = 0$ and (i) follows.

(ii) Let $\alpha_i \in A_i$. Then since each A_i is an ideal in $K[G]$ we have for all $x_1, x_2, \ldots, x_{n-1} \in G$

$$\alpha_1 x_1 \alpha_2 x_2 \cdots \alpha_{n-1} x_{n-1} \alpha_n \in A_1 A_2 \cdots A_n = 0.$$

Thus by (i) above $\theta(\alpha_1)\theta(\alpha_2) \cdots \theta(\alpha_n) = 0$ and this clearly yields $\theta(A_1)\theta(A_2) \cdots \theta(A_n) = 0$. Since $\theta(A_i)$ is an ideal in $K[\Delta]$ by Theorem 1.4, the result follows.

Theorem 20.2 (Passman [*40*]). Let K be a field of characteristic $p > 0$ and let G be a group. Then

(i) $NK[G] = \big(NK[\Delta^p(G)]\big) \cdot K[G]$.

(ii) $NK[\Delta^p(G)] = JK[\Delta^p(G)] = \bigcup JK[W]$ where the union is over all finite normal subgroups W of G contained in $\Delta^p(G)$.

(iii) $NK[G] = 0$ if and only if $\Delta^p(G) = \langle 1 \rangle$.

Proof. Since $\Delta^p(G) \subseteq \Delta^+(G)$, (ii) follows from Lemma 19.4. Moreover by that lemma

$$\big(NK[\Delta^p(G)]\big) \cdot K[G] \subseteq NK[G].$$

We consider the reverse inclusion. Let A be a nilpotent ideal in $K[G]$. Then by Lemma 20.1(ii), $\theta(A)$ is a nilpotent ideal in $K[\Delta]$ so $\theta(A) \subseteq JK[\Delta]$. Since clearly $A \subseteq \theta(A) \cdot K[G]$ and since $NK[G]$ is the sum of all such nilpotent ideals we have

$$NK[G] \subseteq (JK[\Delta]) \cdot K[G].$$

Now Lemma 19.6 applied to $\Delta(G)$ yields

$$JK[\Delta] \subseteq \big(NK[\Delta^p(G)]\big) \cdot K[\Delta]$$

and thus

$$NK[G] \subseteq \big(NK[\Delta^p(G)]\big) \cdot K[G]$$

and (i) follows. Finally (iii) is merely a restatement of Lemma 3.1 and Theorem 3.7.

Theorem 20.3 (Passman [*40*]). Let K be a field of characteristic $p > 0$ and let G be a group. Then the following are equivalent.

(i) $NK[G]$ is nilpotent.

(ii) There exists $\alpha \in K[G]$, $\alpha \neq 0$ with $\alpha NK[G] = 0$.

(iii) $\Delta^p(G)$ is finite.

Proof. (i) $\Rightarrow$ (ii). Let n be minimal with $(NK[G])^n = 0$. Then $(NK[G])^{n-1} \neq 0$ so we may choose $\alpha \in (NK[G])^{n-1}$ with $\alpha \neq 0$. Then $\alpha NK[G] \subseteq (NK[G])^n = 0$.

(ii) $\Rightarrow$ (iii). Since $JK[\Delta^p(G)] \subseteq NK[G]$ by Theorem 20.2, we have $\alpha JK[\Delta^p(G)] = 0$. Write $\alpha = \sum x_i\alpha_i$ where $\alpha_i \in K[\Delta^p(G)]$ and the x_i are in distinct cosets of $\Delta^p(G)$ in G. Then

$$0 = \alpha JK[\Delta^p(G)] = \sum x_i\alpha_i JK[\Delta^p(G)]$$

implies that $\alpha_i JK[\Delta^p(G)] = 0$. Since $\alpha_i \neq 0$ for some i, we have $|\Delta^p(G)| < \infty$ by Lemma 19.5.

(iii) $\Rightarrow$ (i). Since $\Delta^p(G)$ is finite, it follows that $NK[\Delta^p(G)] = JK[\Delta^p(G)]$ is nilpotent. Thus by Lemma 16.7 and Theorem 20.2, $NK[G] = \big(JK[\Delta^p(G)]\big) \cdot K[G]$ is nilpotent.

Corollary 20.4. Let K be a field of characteristic p and let H be a normal subgroup of G. If $NK[G/H] = 0$, then

$$NK[G] \subseteq (NK[H]) \cdot K[G].$$

Proof. By Theorem 20.2, $\Delta^p(G/H) = \langle 1 \rangle$ and hence clearly $\Delta^p(G) \subseteq \Delta^p(H)$. Since $\Delta^p(G)$ is normal in H, Lemma 19.4 and Theorem 20.2 yield

$$NK[G] = \big(JK[\Delta^p(G)]\big) \cdot K[G] \subseteq (NK[H]) \cdot K[G]$$

and the result follows.

Of course if $JK[G]$ is nilpotent, then $JK[G] = NK[G]$ is described as in Theorem 20.2. The following result shows that this condition is in some sense not far from the condition that $K[G]$ be semisimple.

Theorem 20.5. Let K be a field of characteristic p. Then $JK[G]$ is nilpotent if and only if G has subgroups P and H such that

(i) P is a finite p-group and $[G:H] < \infty$.

(ii) P is a normal subgroup of H.

(iii) $K[H/P]$ is semisimple.

Proof. Let P be a finite p-group and let I denote the kernel of the natural map $K[P] \to K[P/P]$. If $\tilde{K}$ is the algebraic closure of K, then by Lemma 10.1(ii) $\tilde{K} \cdot I = J\tilde{K}[P]$ is a nilpotent ideal. Hence I is a nilpotent ideal in $K[P]$ and then clearly $I = JK[P]$. Suppose now that P is normal in some group H. Then $I \cdot K[H]$ is the kernel of the natural map $K[H] \to K[H/P]$ and $I \cdot K[H]$ is a nilpotent ideal by Lemma 16.7. Thus $JK[H/P] = JK[H]/I \cdot K[H]$.

We now proceed to prove the theorem. Suppose first that G has such subgroups P and H satisfying (i), (ii), and (iii) and use the above notation. Since $JK[H/P] = 0$, we conclude that $JK[H] = I \cdot K[H]$ is nilpotent. Thus $JK[G]$ is nilpotent by Lemma 16.8 since $[G:H] < \infty$.

Now assume that $JK[G]$ is nilpotent so that, by Theorem 20.3, $\Delta^p(G)$ is finite. Let P be a Sylow p-subgroup of $\Delta^p(G)$. Since all conjugates of P are contained in $\Delta^p(G)$, we see that $[G:H] < \infty$ where H is the normalizer of P in G. Clearly $P \subseteq \Delta^p(H)$. On the other hand, let $x \in \Delta^p(H)$ have order a power of p. Since $[G:H] < \infty$, it follows that x has only a finite number of conjugates in G and thus $x \in \Delta^p(G)$. But $x \in H$ so x normalizes P and $\langle P, x\rangle$ is a p-subgroup of $\Delta^p(G)$. This implies that $\langle P, x\rangle = P$ so $x \in P$ and thus clearly $P = \Delta^p(H)$. Now by Lemma 16.8, $JK[H]$ is nilpotent so by Theorem 20.2 we have $JK[H] = (JK[P]) \cdot K[H] = I \cdot K[H]$ in the notation of the first paragraph. Thus

$$JK[H/P] = JK[H]/I \cdot K[H] = 0$$

and the result follows.

We now come to an interesting application of the above. A ring R is said to be perfect (see [*8*]) if

(1) R/JR is Artinian.

(2) For every sequence $\alpha_1, \alpha_2, \ldots$ of elements of JR there exists an integer m with $\alpha_1\alpha_2 \cdots \alpha_m = 0$.

The question of when $K[G]$ is perfect is answered below.

Lemma 20.6. Let R be a perfect ring. Then

(i) There exists $\alpha \in R$, $\alpha \neq 0$ with $\alpha JR = 0$.

(ii) If NR is nilpotent, then $JR = NR$.

Proof. If $JR = 0$, then the result is trivial so we will assume that $JR \neq 0$. Let $A = \{\alpha \in JR \mid \alpha JR = 0\}$. Suppose $A = 0$. We now construct an infinite sequence $\alpha_1, \alpha_2, \ldots$ of elements of JR for which $\alpha_1\alpha_2 \cdots \alpha_n \neq 0$ for all n. First $JR \neq 0$ so let $\alpha_1 \in JR$, $\alpha_1 \neq 0$. Suppose that we have already found $\alpha_1, \alpha_2, \ldots, \alpha_n \in JR$ with $\beta = \alpha_1\alpha_2 \cdots \alpha_n \neq 0$. Then

$\beta \in JR$ and $\beta \notin A$ so $\beta JR \neq 0$. Thus there exists $\alpha_{n+1} \in JR$ with $\alpha_1\alpha_2 \cdots \alpha_n\alpha_{n+1} = \beta\alpha_{n+1} \neq 0$ and we let α_{n+1} be the next element in the sequence. Since the existence of such a sequence contradicts the fact that R is perfect, we conclude that $A \neq 0$.

Let $\alpha \in A$, $\alpha \neq 0$. Then $\alpha JR = 0$ so (i) follows. Set $I = R\alpha R$, the ideal of R generated by α. Then $I \subseteq JR$ so $I^2 \subseteq R \cdot \alpha JR = 0$ and therefore $NR \neq 0$. Suppose now that NR is nilpotent and let $\bar{R} = R/NR$. Then $J\bar{R} = JR/NR$ so clearly $\bar{R}$ is also perfect. If $\bar{B}$ is any nilpotent ideal in $\bar{R}$, then since NR is nilpotent it follows that B, the complete inverse image of $\bar{B}$ in R, is also nilpotent. Hence $B \subseteq NR$, so $\bar{B} = 0$ and thus $N\bar{R} = 0$. The above argument applied to $\bar{R}$ implies that $J\bar{R} = 0$ so that $JR = NR$ and the lemma is proved.

Theorem 20.7 (Woods [*61*]). The group ring $K[G]$ is perfect if and only if G is finite.

Proof. If G is finite, then $K[G]$ is Artinian and $JK[G]$ is nilpotent. Thus $K[G]$ is clearly perfect.

Suppose now that $K[G]$ is perfect. If K has characteristic 0, then $NK[G] = 0$ by Lemma 3.1 and Theorem 3.3 so by Lemma 20.6(ii) we have $JK[G] = 0$. Thus $K[G]$ is Artinian and G is finite by Theorem 2.6. Now let K have characteristic p. Then by Lemma 20.6(i) there exists $\alpha \in K[G]$, $\alpha \neq 0$ with $\alpha NK[G] \subseteq \alpha JK[G] = 0$. Thus Theorem 20.3 implies that $\Delta^p(G)$ is finite and $NK[G]$ is nilpotent and by Lemma 20.6(ii) we have $JK[G] = NK[G]$. Let I denote the kernel of the natural map $K[\Delta^p] \to K[\Delta^p/\Delta^p]$. Then I is a maximal right ideal of $K[\Delta^p]$, so $I \supseteq JK[\Delta^p]$. Now $I \cdot K[G]$ is clearly the kernel of the natural map $K[G] \to K[G/\Delta^p]$ and

$$I \cdot K[G] \supseteq (JK[\Delta^p]) \cdot K[G] = NK[G]$$

by Theorem 20.2. Since $JK[G] = NK[G]$, it then follows that $K[G/\Delta^p]$ is a homomorphic image of the Artinian ring $K[G]/JK[G]$. Thus $K[G/\Delta^p]$ is Artinian and G/Δ^p is finite by Theorem 2.6. Since Δ^p is also finite, we have $|G| < \infty$ and the result follows.

§21. MORE EXAMPLES

In this section we offer a number of interesting examples. These are most conveniently given in terms of wreath products. Let G and H be abstract groups. Then the wreath product $G \wr H$ of G by H is described as follows. For each $x \in H$, we let G_x be the set of ordered pairs

$$G_x = \{[g, x] \mid g \in G\}$$

with multiplication defined by

$$[g_1, x][g_2, x] = [g_1g_2, x].$$

Thus in this way G_x is clearly a group isomorphic to G. If $y \in H$ then y induces an isomorphism $G_x \to G_{xy}$ by $[g, x] \to [g, xy]$ and thus y clearly induces an automorphism of

$$W = \prod_{x \in H} G_x,$$

the direct product of the groups G_x. This yields an action of H on W and we set $G \wr H = W \times_\sigma H$, the semidirect product of W by H. Roughly speaking, $G \wr H$ has a normal subgroup $\bar{W} \simeq W$ which is a direct product of copies of G indexed by the elements of H and a complementary subgroup $\bar{H} \simeq H$ such that $G \wr H = \bar{W}\bar{H}$ and $\bar{H}$ acts on $\bar{W}$ as H does on W.

Lemma 21.1. Let G be a group with a normal abelian subgroup A and let $H = \{x \in G \mid [A : \mathbf{C}_A(x)] < \infty\}$. Then H is a normal subgroup of G containing A. Let K be a field and let I be a nonzero ideal in $K[G]$. Then $I \cap K[H] \neq 0$.

Proof. Since A is a normal abelian subgroup of G, it follows immediately that H is a normal subgroup of G and $H \supseteq A$.

Since $I \neq 0$, there exists $\alpha \in I$ with $1 \in \operatorname{Supp} \alpha$ and thus $I_0 = \{\alpha \in I \mid 1 \in \operatorname{Supp} \alpha\}$ is nonempty. For each $\alpha \in I_0$, let $N(\alpha)$ denote the number of cosets of A in $(\operatorname{Supp} \alpha)A - H$ so that $N(\alpha) \geq 0$. Now suppose $\alpha \in I_0$ is chosen so that $N(\alpha)$ is minimal and write $\alpha = \sum_1^n x_i\alpha_i$ where $\alpha_i \in K[A]$, $\alpha_i \neq 0$, and the x_i are in distinct cosets of A in G. We may suppose the notation so chosen that $x_1, x_2, \ldots, x_r \in H$ and $x_{r+1}, x_{r+2}, \ldots, x_n \in G - H$. Moreover we may take $x_1 = 1$ since $\alpha \in I_0$ implies that $1 \in \operatorname{Supp} \alpha$. By definition $N(\alpha) = n - r$. If $N(\alpha) = 0$, then $\alpha \in I \cap K[H]$ and the result follows. We assume now that $N(\alpha) \neq 0$ and derive a contradiction.

Since $1 \in \operatorname{Supp} \alpha_1$, there are only a finite number of $b \in A$ with $1 \notin \operatorname{Supp} \alpha_1(b - 1)$ and since $x_n \in G - H$ there are an infinite number of distinct commutators $(x_n, a) = x_n^{-1}a^{-1}x_na$ for $a \in A$. Hence we can choose $a \in A$ so that if $b = (x_n, a)$, then $1 \in \operatorname{Supp} \alpha_1(b - 1)$. Set

$$\beta = \alpha b - a^{-1}\alpha a = \sum_1^n x_i\beta_i.$$

Then since A is normal and abelian, we have

$$\beta_i = \alpha_i(b - (x_i, a)) \in K[A].$$

Thus $\beta_n = 0$ and $\beta_1 = \alpha_1(b - 1)$ so $1 \in \operatorname{Supp} \beta_1$. Since I is an ideal we then have $\beta \in I_0$ and clearly $N(\beta) < n - r = N(\alpha)$, a contradiction. Thus $N(\alpha) = 0$ and $I \cap K[H] \neq 0$.

Theorem 21.2. Any of the following conditions imply that the group ring $K[G]$ is semisimple.

(i) G has a normal abelian subgroup A with $JK[A] = 0$ and such that if $x \in G$ with $[A:\mathbf{C}_A(x)] < \infty$, then $x \in A$.

(ii) $G = A \wr B$ where A is a nonidentity abelian group with $JK[A] = 0$ and B is infinite.

(iii) G has a normal torsion free abelian group A with $A = \mathbf{C}_G(A)$.

(iv) $G = A \wr B$ where A is a nonidentity torsion free abelian group.

Proof. (i) In the notation of Lemma 21.1 we must have $H = A$. Suppose now that $I = JK[G] \neq 0$. Then by that lemma, $JK[G] \cap K[A] = I \cap K[A] \neq 0$. Since $JK[G] \cap K[A] \subseteq JK[A] = 0$ by Lemma 16.9, we have a contradiction.

(ii) Since $G = A \wr B$, it follows that G has a normal abelian subgroup $W = \prod_{b\in B} A_b$ which is a direct product of copies of A indexed by the elements of B and $G/W \simeq B$. If K has characteristic 0, then $JK[W] = 0$ by Corollary 17.8. If K has characteristic p, then since $JK[A] = 0$ we conclude that A has no elements of order p. Thus W has no elements of order p and again $JK[W] = 0$ by Corollary 17.8. Now $G/W \simeq B$ acts on W and if $x \in B$, $x \neq 1$, then clearly $\mathbf{C}_W(x)$ consists of all those elements of W whose projections into the factors A_b are constant on the orbits of $\langle x\rangle$. Since $A \neq \langle 1\rangle$ and B is infinite, this implies easily that $[W:\mathbf{C}_W(x)] = \infty$. The result follows from (i).

(iii) Since A is torsion free abelian we have $JK[A] = 0$ by Corollary 17.8. Suppose $x \in G$ with $[A:\mathbf{C}_A(x)] = n < \infty$. If $a \in A$, then $a^n \in \mathbf{C}_A(x)$ so since A is abelian we have

$$(a^x a^{-1})^n = (a^n)^x a^{-n} = 1.$$

Hence $a^x = a$ since A is torsion free and $x \in \mathbf{C}_G(A) = A$. The result follows from (i).

(iv) Since $G = A \wr B$, it follows that G has a normal subgroup $W = \prod_{b\in B} A_b$. Certainly W is torsion free abelian. Now $G/W \simeq B$ acts on W and if $x \in B$, $x \neq 1$, then clearly $\mathbf{C}_W(x)$ consists of all elements of W whose projections into the factors A_b are constant on the orbits of $\langle x\rangle$.

Then $A \neq \langle 1 \rangle$ yields $\mathbf{C}_W(x) \neq W$ so $\mathbf{C}_G(W) = W$. The result follows from (iii).

Suppose for example that K has characteristic p and that $G = Z \wr Z_p$ where Z is infinite cyclic and Z_p is cyclic of order p. Then G has infinitely many elements of order p and yet by (iv) above $JK[G] = 0$. Thus an exact converse to either Theorem 3.6 or Theorem 18.4 is decidedly false.

Lemma 21.3. Let H be a normal subgroup of G and let $\alpha \in JK[G] \cap K[H]$. Let $x \in G$ map to an element of infinite order in G/H. Then for some integer $n \geq 0$ we have

$$\alpha\alpha^x\alpha^{x^2} \cdots \alpha^{x^n} = 0.$$

Proof. Let $G_1 = \langle H, x \rangle$. Then by Lemma 16.9 $\alpha x^{-1} \in JK[G] \cap K[G_1] \subseteq JK[G_1]$. Let $\beta = \sum \beta_i x^i$ be a quasi right inverse for αx^{-1} in $K[G_1]$ with $\beta_i \in K[H]$. Then $(\alpha x^{-1}) + \beta + (\alpha x^{-1})\beta = 0$ yields

$$\alpha x^{-1} + \sum \beta_i x^i + \sum \alpha\beta_i^x x^{i-1} = 0.$$

Hence since x has infinite order in G/H, we have

$$\beta_i = -\alpha\beta_{i+1}^x \quad \text{for} \quad i \neq -1$$

$$\beta_{-1} = -\alpha - \alpha\beta_0^x.$$

Now $\beta_q = 0$ for some $q > 0$, and thus the first equation above and induction imply that $\beta_i = 0$ for all $q \geq i \geq 0$. Then $\beta_{-1} = -\alpha$ and the first equation and induction yield

$$\beta_{-i} = (-1)^i\alpha\alpha^x \cdots \alpha^{x^{i-1}}.$$

Since $\beta_{-(n+1)} = 0$ for some $n \geq 0$, the result follows.

Theorem 21.4. Let $G = A \wr B$ where B is abelian and has no elements of order p in case K has characteristic p. If B contains an element of infinite order, then $JK[G] = 0$.

Proof. G has a normal subgroup $W = \prod_{b \in B} A_b$ which is the direct product of copies of A indexed by the elements of B. Moreover $G/W \simeq B$. By Theorem 17.7 it suffices to show that $JK[G] \cap K[W] = 0$. Let $\alpha \in JK[G] \cap K[W]$. Then there exists $b_1, b_2, \ldots, b_m \in B$ such that $\alpha \in K[H]$ where $H = A_{b_1} \times A_{b_2} \times \cdots \times A_{b_m}$. Let $y \in B$ have infinite order and think of y as being an element of G. Choose integer r sufficiently large so that the elements $b_1, b_2, \ldots, b_m$ are in distinct cosets of $\langle y^r \rangle$ in B. Then $x = y^r$ has infinite order in G/W and by Lemma 21.3 there exists

$n \geq 0$ such that $\alpha\alpha^x \cdots \alpha^{x^n} = 0$. Now $\alpha^{x^i} \in K[H^{x^i}]$ and by our choice of x, $HH^x \cdots H^{x^n} = H \times H^x \times \cdots \times H^{x^n}$ is a direct product. This clearly yields $\alpha = 0$ and the result follows.

As an example, take $G = Z_p \wr Z$ and let K have characteristic p. Then G has a nontrivial normal abelian p-subgroup and yet $JK[G] = 0$ by the above. We now consider infinite p-groups.

Lemma 21.5. Let G, A, and B be groups.

(i) If G is a locally finite p-group and if K is a field of characteristic p, then

$$JK[G] = \{\sum a_x \cdot x \mid \sum a_x = 0\}.$$

(ii) If A and B are p-groups, then so is $A \wr B$.

(iii) If A and B are locally finite, then so is $A \wr B$.

(iv) If $A \neq \langle 1 \rangle$ and if B is infinite, then $A \wr B$ has no nontrivial finite normal subgroups.

Proof. (i) Let $I = \{\sum a_x \cdot x \in K[G] \mid \sum a_x = 0\}$. Then I is a maximal right ideal in $K[G]$ and so $I \supseteq JK[G]$. Let $\alpha \in I$. Since G is locally finite, there exists a finite subgroup H of G such that $\alpha \in K[H]$. Let $\tilde{K}$ denote the algebraic closure of K. Then $\alpha \in I$, $\alpha \in \tilde{K}[H]$, and H is a finite p-group so $\alpha \in J\tilde{K}[H]$ by Lemma 10.1(ii) and hence α is nilpotent. Thus I is a nil ideal, $I \subseteq JK[G]$, and (i) follows.

Now $A \wr B$ has a normal subgroup $W = \prod_{b \in B} A_b$, and $(A \wr B)/W \simeq B$.

(ii) This is obvious since W is also a p-group.

(iii) Let L be a finitely generated subgroup of $A \wr B$. Then LW/W is a finitely generated subgroup of B and thus by assumption $[L:L \cap W] = [LW:W] < \infty$. Since L is finitely generated so is $L \cap W$ by Lemma 6.1 and since W is clearly locally finite we have $|L \cap W| < \infty$. Thus L is finite.

(iv) Let L be a finite normal subgroup of $A \wr B$ and let $L_0 = L \cap W$ so that L_0 is also a normal subgroup of $A \wr B$. Suppose that $L_0 \neq \langle 1 \rangle$ and choose a minimal set $\{b_1, b_2, \ldots, b_n\}$ of elements of B such that $L_0 \subseteq A_{b_1} \times A_{b_2} \times \cdots \times A_{b_m}$. It follows then that $\{b_1, b_2, \ldots, b_m\}$ is a union of orbits of the regular permutation representation of B, a contradiction since B is infinite. Thus $L \cap W = L_0 = \langle 1 \rangle$ and L centralizes W. Now $A \wr B$ acts by conjugation on the set of subgroups $\{A_b \mid b \in B\}$ and the kernel of this action contains W. Since $A \neq \langle 1 \rangle$, this easily implies that the kernel is precisely W. Therefore we must have $L \subseteq W$ and hence $L = \langle 1 \rangle$.

Theorem 21.6. Let A and B be locally finite p-groups with $A \neq \langle 1 \rangle$ and with B infinite. If $G = A \wr B$ and if K is a field of characteristic p, then

$$JK[G] = \{\sum a_x \cdot x \mid \sum a_x = 0\}$$

and $NK[G] = 0$.

Proof. The form of $JK[G]$ follows from (i), (ii), and (iii) of the previous lemma. By Lemma 21.5(iv) and Theorem 3.7 we have $NK[G] = 0$.

As an example take $A = Z_p$ and $B = \prod Z_p$, an infinite direct product of copies of Z_p. Then for K a field of characteristic p, we have $JK[A \wr B] \neq 0$ and $NK[A \wr B] = 0$. Hence for arbitrary groups G there is in general little relation between $JK[G]$ and $NK[G]$.

IV

IDEMPOTENTS AND ANNIHILATORS

§22. TRACE OF IDEMPOTENTS

In this section K will be a field of characteristic 0 and C will denote the field of complex numbers. If

$$\alpha = \sum_{x \in G} a_x \cdot x \in K[G]$$

then we define the trace of α to be $\operatorname{tr} \alpha = a_1$, the coefficient of $1 \in G$ in α.

Lemma 22.1. The map $\operatorname{tr}: K[G] \to K$ is K-linear and for $\alpha, \beta \in K[G]$ we have $\operatorname{tr} \alpha\beta = \operatorname{tr} \beta\alpha$.

Proof. Certainly tr is K-linear. If $\alpha = \sum a_x \cdot x$ and $\beta = \sum b_x \cdot x$, then

$$\operatorname{tr} \alpha\beta = \sum_x a_x b_{x^{-1}}$$

and this is symmetric in the a's and b's.

We will concern ourselves here with the trace of idempotents in $K[G]$. To start with we have

Lemma 22.2. Let $e \in K[G]$ be an idempotent and suppose that G is finite. Then $\operatorname{tr} e = r/|G|$ where $r = \dim_K K[G]e$. In particular $\operatorname{tr} e$ is rational and $0 < \operatorname{tr} e \leq 1$.

Proof. View $V = K[G]$ as a right $K[G]$-module. Then with respect to the natural basis $\{x \mid x \in G\}$ every element of $K[G]$ can be written as a matrix and we let $s(\alpha)$ denote the trace of the matrix of $\alpha \in K[G]$. It then follows easily that for $x \in G$, $s(x) = 0$ if $x \neq 1$ and $s(1) = |G|$. Since s is K-linear, if $\alpha = \sum a_x \cdot x$ is an element of $K[G]$, then we have

$$s(\alpha) = a_1 s(1) = (\text{tr } \alpha)\, |G|.$$

Now of course $s(\alpha)$ is independent of the choice of basis for V. If e is an idempotent, then $V = K[G]e + K[G](1 - e)$ yields a vector space direct sum decomposition. Since e acts like 1 on $K[G]e$ and like 0 on $K[G](1 - e)$, it follows that

$$s(e) = \dim K[G]e = r$$

and the lemma is proved.

Thus the case of finite groups is fairly simple. We now consider the infinite case and we start by studying $C[G]$. Let $\alpha = \sum a_x \cdot x, \beta = \sum b_x \cdot x$, $\gamma = \sum c_x \cdot x$ be elements of $C[G]$. We define an inner product and appropriate norms on $C[G]$ by

$$(\alpha, \beta) = \sum_x a_x \bar{b}_x$$

$$\|\alpha\| = (\alpha, \alpha)^{\frac{1}{2}} = \left(\sum_x |a_x|^2\right)^{\frac{1}{2}}$$

$$|\alpha| = \sum_x |a_x|$$

where the overbar denotes complex conjugation and $|a|$ is the absolute value of $a \in C$. If L is a C-subspace of $C[G]$, then we define the distance between L and γ to be

$$d(L, \gamma) = \inf_{\alpha \in L} \|\alpha - \gamma\|.$$

The following lemma holds in any inner product space. The one after indicates properties which are more dependent on the group algebra structure.

Lemma 22.3. Let $\alpha, \beta \in L$, a linear subspace of $C[G]$. Then

$$|(\beta, \alpha - \gamma)|^2 \leq \|\beta\|^2(\|\alpha - \gamma\|^2 - d(L, \gamma)^2).$$

Proof. This is trivial for $\beta = 0$ so assume $\beta \neq 0$ and set $k = (\alpha - \gamma, \beta)/\|\beta\|^2$. Then $\alpha - k\beta \in L$ so

$$\|\alpha - k\beta - \gamma\|^2 \geq d(L, \gamma)^2.$$

Thus

$$
\begin{aligned}
\|\alpha - \gamma\|^2 - d(L, \gamma)^2 \geq \|\alpha - \gamma\|^2 - \|\alpha - k\beta - \gamma\|^2 \\
&= (\alpha - \gamma, \alpha - \gamma) - (\alpha - \gamma - k\beta, \alpha - \gamma - k\beta) \\
&= k(\beta, \alpha - \gamma) + \bar{k}(\alpha - \gamma, \beta) - k\bar{k}(\beta, \beta) \\
&= |(\beta, \alpha - \gamma)|^2/\|\beta\|^2
\end{aligned}
$$

and the result follows.

Lemma 22.4. With the above notation, we have

(i) $\|\alpha + \beta\| \leq \|\alpha\| + \|\beta\|$.

(ii) $|\text{tr } \alpha| \leq \|\alpha\|$, $(\alpha, 1) = \text{tr } \alpha$.

(iii) $(\alpha, \beta\gamma) = (\alpha\gamma^*, \beta)$ where $\gamma^* = \sum \bar{c}_x x^{-1}$.

(iv) $\|\alpha\beta\| \leq \|\alpha\| \cdot |\beta|$.

Proof. (i) This is just the triangle inequality. Set $L = C[G]$ and $\gamma = 0$ in Lemma 22.3. Then $d(C[G], 0) = 0$ so for all $\alpha, \beta \in C[G]$ we have by that lemma, $|(\beta, \alpha)| \leq \|\beta\| \cdot \|\alpha\|$. Therefore,

$$
\begin{aligned}
\|\alpha + \beta\|^2 &= (\alpha + \beta, \alpha + \beta) \\
&= \|\alpha\|^2 + (\alpha, \beta) + (\beta, \alpha) + \|\beta\|^2 \\
&\leq \|\alpha\|^2 + 2\,\|\alpha\| \cdot \|\beta\| + \|\beta\|^2 \\
&= (\|\alpha\| + \|\beta\|)^2
\end{aligned}
$$

and (i) follows. Part (ii) is obvious.

(iii) Now it is easy to see that the map $*: \gamma \to \gamma^*$ is in fact an anti-automorphism of $C[G]$. Moreover

$$(\alpha, \beta) = \sum_x a_x \bar{b}_x = \text{tr } \alpha\beta^*.$$

Thus

$$
\begin{aligned}
(\alpha, \beta\gamma) &= \text{tr } \alpha(\beta\gamma)^* \\
&= \text{tr } \alpha\gamma^*\beta^* = (\alpha\gamma^*, \beta)
\end{aligned}
$$

and (iii) is proved. As a special case we observe that for $x \in G$

$$
\begin{aligned}
\|\alpha x\| &= (\alpha x, \alpha x)^{\frac{1}{2}} = (\alpha x x^*, \alpha)^{\frac{1}{2}} \\
&= (\alpha, \alpha)^{\frac{1}{2}} = \|\alpha\|
\end{aligned}
$$

since $x^* = x^{-1}$.

(iv) Finally by (i) and the above

$$\begin{aligned}\|\alpha\beta\| &= \sum_x \alpha b_x \cdot x \le \sum_x \|\alpha b_x \cdot x\| \\ &= \sum_x \|\alpha\| \cdot |b_x| = \|\alpha\| \cdot |\beta|\end{aligned}$$

and the result follows.

Now let $e \neq 0$ be an idempotent in $C[G]$ and set $M = eC[G]$. Then M is a linear subspace of $C[G]$ and we let $d = d(M, 1)$ be the distance between M and 1. For each integer $n > 0$, choose $\alpha_n \in M$ with

$$\|\alpha_n - 1\|^2 < d^2 + 1/n^4.$$

Lemma 22.5. There exist nonnegative real constants r' and r'' with

(i) $| \, \|\alpha_n\|^2 - \operatorname{tr} \alpha_n| \le r'/n$.

(ii) $\|\alpha_n e - e\| \le r''/n$.

Proof. By Lemma 22.3 with $L = M$, $\gamma = 1$, we have for all $\beta \in M$

$$|(\beta, \alpha_n - 1)| \le \|\beta\|/n^2.$$

We observe also that

$$\|\alpha_n - 1\| \le (d^2 + 1/n^4)^{\frac{1}{2}} \le d + 1$$

and by Lemma 22.4(i)

$$\|\alpha_n\| \le \|1\| + \|\alpha_n - 1\| \le d + 2.$$

(i) Since $\alpha_n \in M$, the above yields

$$\begin{aligned}|(\alpha_n, \alpha_n - 1)| &\le \|\alpha_n\|/n^2 \\ &\le \|\alpha_n\|/n \le (d + 2)/n.\end{aligned}$$

Moreover by Lemma 22.4(ii) we have

$$\begin{aligned}(\alpha_n, \alpha_n - 1) &= (\alpha_n, \alpha_n) - (\alpha_n, 1) \\ &= \|\alpha_n\|^2 - \operatorname{tr} \alpha_n\end{aligned}$$

so (i) follows with $r' = d + 2$.

(iii) By Lemma 22.4(iii)

$$\begin{aligned}\|\alpha_n e - e\|^2 &= \big((\alpha_n - 1)e, (\alpha_n - 1)e\big) \\ &= ((\alpha_n - 1)ee^*, \alpha_n - 1).\end{aligned}$$

Now $(\alpha_n - 1)ee^* = \alpha_n ee^* - ee^* \in M$ so the above and Lemma 22.4(iv) yield

$$\begin{aligned}\|\alpha_n e - e\|^2 &\leq \|(\alpha_n - 1)ee^*\|/n^2 \\ &\leq \|\alpha_n - 1\| \cdot |ee^*|/n^2 \\ &\leq (d+1) \cdot |ee^*|/n^2.\end{aligned}$$

Thus the result follows with $(r'')^2 = (d+1) \cdot |ee^*|$.

Lemma 22.6. The trace of e is real and in fact $\operatorname{tr} e \geq \|e\|^2/|e|^2 > 0$.

Proof. By Lemma 22.4(ii) and Lemma 22.5 we have

$$| \, \|\alpha_n\|^2 - \operatorname{tr} \alpha_n| \leq r'/n$$
$$|\operatorname{tr} \alpha_n e - \operatorname{tr} e| \leq \|\alpha_n e - e\| \leq r''/n.$$

Moreover by Lemma 22.1

$$\operatorname{tr} \alpha_n e = \operatorname{tr} e\alpha_n = \operatorname{tr} \alpha_n$$

since $\alpha_n \in M$ implies that $e\alpha_n = \alpha_n$. Thus we have from the above

$$| \, \|\alpha_n\|^2 - \operatorname{tr} e| \leq (r' + r'')/n$$

and we conclude that

$$\operatorname{tr} e = \lim_{n \to \infty} \|\alpha_n\|^2.$$

Therefore $\operatorname{tr} e$ is real and nonnegative.

Now Lemma 22.4(i), (iv) and Lemma 22.5(ii) yield

$$\|e\| \leq \|e - \alpha_n e\| + \|\alpha_n e\| \leq r''/n + \|\alpha_n\| \cdot |e|.$$

Thus taking limits as $n \to \infty$, we obtain

$$\|e\| \leq (\operatorname{tr} e)^{\frac{1}{2}} \, |e| \qquad \text{and} \qquad \operatorname{tr} e \geq \|e\|^2/|e|^2 > 0.$$

We now come to the main theorem of this section. Amazingly enough, an analytic proof yields an algebraic result here.

Theorem 22.7 (Kaplansky [*28*]). Let K be a field of characteristic 0 and let $e \neq 0, 1$ be an idempotent in $K[G]$. Then $\operatorname{tr} e$ is a totally real algebraic number with the property that it and all its algebraic conjugates lie strictly between 0 and 1.

Proof. Since $\operatorname{Supp} e$ is finite, there exists a finitely generated field extension K_0 of the rationals such that $K_0 \subseteq K$ and $e \in K_0[G]$. Now K_0 is embeddable in the complex numbers C so we can view e as an element of $C[G]$. Certainly e is an idempotent in $C[G]$ and $e \neq 0, 1$.

By Lemma 22.6 we have $\operatorname{tr} e > 0$. Since $1 - e$ is also a nonzero idempotent in $C[G]$, Lemma 22.6 yields $1 - \operatorname{tr} e = \operatorname{tr}(1 - e) > 0$ so $\operatorname{tr} e < 1$. Let σ be a field automorphism of the complex numbers. Then σ clearly induces a ring automorphism of $C[G]$ by

$$\alpha = \sum_x a_x \cdot x \to \alpha^\sigma = \sum_x (a_x)^\sigma \cdot x.$$

Since e^σ is again an idempotent in $C[G]$ and $\operatorname{tr} e^\sigma = (\operatorname{tr} e)^\sigma$, the above yields $1 > (\operatorname{tr} e)^\sigma > 0$ for all σ. Now if $\operatorname{tr} e$ is transcendental over the rationals, then there certainly exists a field automorphism σ such that $(\operatorname{tr} e)^\sigma$ is not real, a contradiction. Thus $\operatorname{tr} e$ is algebraic and the theorem is proved.

There is an interesting consequence of this result which we offer below.

Corollary 22.8 (Kaplansky [*28*]). Let K be a field of characteristic 0 and let $\alpha, \beta \in K[G]$. If $\alpha\beta = 1$, then $\beta\alpha = 1$.

Proof. Suppose $\alpha\beta = 1$ and set $e = \beta\alpha$. Then

$$e^2 = \beta(\alpha\beta)\alpha = \beta\alpha = e$$

so e is an idempotent in $K[G]$. Moreover by Lemma 22.1

$$\operatorname{tr} e = \operatorname{tr} \beta\alpha = \operatorname{tr} \alpha\beta = 1.$$

Thus by the previous theorem we must have $e = 1$.

§23. CENTRAL IDEMPOTENTS

A good deal more can be said about idempotents which are central in group rings. We start by considering finite groups in characteristic $p > 0$.

Theorem 23.1 (Brauer [*10*]). Let G be a finite group, P a p-subgroup of G, and let K be a field of characteristic p. Then the natural projection $\pi: K[G] \to K[\mathbf{C}(P)]$ induces a ring homomorphism, the Brauer homomorphism, from the center of $K[G]$ to the center of $K[\mathbf{C}(P)]$.

Proof. Let $\alpha = \sum a_x \cdot x$, $\beta = \sum b_x \cdot x$ be central elements in $K[G]$. Since π is clearly a K-linear map, we must show that $\pi(\alpha)$ is central in $K[\mathbf{C}(P)]$ and that $\pi(\alpha\beta) = \pi(\alpha)\pi(\beta)$.

Let $w \in G$. Then since α is central in $K[G]$ we have

$$\sum_x a_x \cdot (w^{-1}xw) = w^{-1}\alpha w = \alpha = \sum_x a_x \cdot x$$

so that for all $x \in G$, $a_{w^{-1}xw} = a_x$. This implies that the coefficients a_x of α are constant on the conjugacy classes of G and hence clearly the coefficients of $\pi(\alpha)$ are constant on the conjugacy classes of $\mathbf{C}(P)$. Thus $\pi(\alpha)$ is central in $K[\mathbf{C}(P)]$.

Write $\alpha\beta = \sum c_x \cdot x$ so that

$$c_x = \sum_{yz=x} a_y b_z.$$

Fix $x \in \mathbf{C}(P)$ and let $\mathcal{S} = \{((y, z)) \mid yz = x\}$, where $((y, z))$ designates an ordered pair. If $h \in P$ and if $((y, z)) \in \mathcal{S}$, then

$$y^h z^h = (yz)^h = x^h = x$$

since $x \in \mathbf{C}(P)$. Then $((y^h, z^h)) \in \mathcal{S}$ and in this way P acts as a permutation group on $\mathcal{S}$. Let the orbits of this action be $\mathcal{S}_1, \mathcal{S}_2, \ldots, \mathcal{S}_k$ and say $((y_i, z_i)) \in \mathcal{S}_i$. Since both α and β are central in $K[G]$, it follows that if $((y, z)) \in \mathcal{S}_i$, then $((y, z)) = ((y_i^h, z_i^h))$ for some $h \in P$ and hence

$$a_y b_z = a_{h^{-1}y_i h} b_{h^{-1}z_i h} = a_{y_i} b_{z_i}.$$

This yields

$$c_x = \sum_{i=1}^{k} |\mathcal{S}_i| \, a_{y_i} b_{z_i}.$$

Now P is a p-group so all orbits have size a power of p. If $p \mid |\mathcal{S}_i|$ then since K has characteristic p, $|\mathcal{S}_i| = 0$ in K and this term can be deleted from the above sum. On the other hand $|\mathcal{S}_i| = 1$ if and only if y_i, $z_i \in \mathbf{C}(P)$. Thus we have

$$c_x = \sum_{\substack{yz=x \\ y,z\in\mathbf{C}(P)}} a_y b_z$$

and this is the coefficient of x in $\pi(\alpha)\pi(\beta)$. Since this is true for all $x \in \mathbf{C}(P)$, we therefore have $\pi(\alpha\beta) = \pi(\alpha)\pi(\beta)$ and the result follows.

Theorem 23.2 (Osima [*34*]). *Let G be a finite group and let K, be a field of characteristic p. If e is a central idempotent in $K[G]$, then* Supp e *consists of p'-elements of G, that is, elements whose order is prime to p.*

Proof. Let $S = [K[G], K[G]]$ be the set of all finite sums of Lie products $[\alpha, \beta] = \alpha\beta - \beta\alpha$ with $\alpha, \beta \in K[G]$. Clearly S is spanned over K by all Lie products $[y, z]$ with $y, z \in G$. We now observe that if $\beta \in S$, then Supp β contains no central elements of G. For suppose that x is a central element of G and $x \in$ Supp β. Then for some $y, z \in G$ we must have $x \in$ Supp$[y, z]$. But this implies that y and z commute so $[y, z] = 0$, a contradiction.

Suppose now that z is an element of Supp e which is not a p'-element and write $z = xy = yx$ where $x \neq 1$ has order a power of p and where q, the order of y, is prime to p. Let P be the p-group $P = \langle x \rangle$. Then by Theorem 23.1, $\pi(e)$ is a central idempotent in $K[\mathbf{C}(P)]$ and $z \in \operatorname{Supp} \pi(e)$. Thus it clearly suffices to assume that x is central in G.

Choose integer n with $p^n \geq |G|$ and with $p^n \equiv 1 \pmod q$ and set $\alpha = y^{-1}e$. If $\alpha = \sum a_g \cdot g$, then by Lemma 3.4

$$\alpha^{p^n} \equiv \sum (a_g)^{p^n} \cdot g^{p^n} \mod S.$$

Now $p^n \geq |G|$ so g^{p^n} is a p'-element and hence by the observation of the previous paragraph and the fact that $x \neq 1$ is central, we have $x \notin \operatorname{Supp} \alpha^{p^n}$. On the other hand, since e is a central idempotent and since $p^n \equiv 1 \pmod q$

$$\alpha^{p^n} = (y^{-1})^{p^n} e^{p^n} = y^{-1}e = \alpha.$$

Finally by the definition of $\alpha = y^{-1}e$, we have $x = y^{-1}z \in \operatorname{Supp} \alpha$, a contradiction, and the result follows.

Let θ^+ denote the natural projection map $\theta^+ : K[G] \to K[\Delta^+(G)]$. Part (ii) of the next lemma will be needed in our later study of annihilator ideals.

Lemma 23.3. Let $H = \Delta(H)$ be a finitely generated group and let K be an algebraically closed field.

(i) If e is a central idempotent in $K[H]$, then $e \in K[\Delta^+(H)]$.

(ii) Suppose that $K[H]$ is semiprime. If I is an ideal in $K[H]$, then there exists a central element $\zeta \neq 0$ in $K[H]$ which is not a zero divisor in $K[H]$ such that $\zeta\theta^+(I) \subseteq I$.

Proof. By Lemma 2.2, $\Delta^+(H)$ is finite and H/Δ^+ is a finitely generated torsion free abelian group. Moreover $[H:\mathbf{Z}(H)] < \infty$. By Lemma 6.1, $\mathbf{Z}(H)$ is finitely generated and hence $\mathbf{Z}(H)$ contains a torsion free abelian subgroup A of finite index.

Set $Z = K[A]$ so that Z is a central subalgebra of $K[H]$. Moreover by Lemma 2.4 no nonzero element of Z is a zero divisor in $K[H]$ since A is torsion free abelian. It is then trivial to form the ring of quotients $E = Z^{-1}K[H]$. This is the set of all formal fractions $\eta^{-1}\alpha$ with $\eta \in Z$, $\eta \neq 0$, $\alpha \in K[H]$, and with the usual identifications made. If $F = Z^{-1}K[A]$, then F is certainly a central subfield of E and since $[H:A] < \infty$, E is a finite dimensional F-algebra.

Let $\mathcal{G} = \operatorname{Hom}(H/\Delta^+, K^*)$. Then as in Section 17, $\mathcal{G}$ acts as a group of algebra automorphisms of $K[H]$. In fact if $\lambda \in \operatorname{Hom}(H/\Delta^+, K^*)$, then λ

induces naturally a multiplicative homomorphism $H \to K^*$ and we define $\lambda^*: K[H] \to K[H]$ by

$$\lambda^*(\sum a_x \cdot x) = \sum a_x \lambda(x) \cdot x.$$

Certainly for λ, $\mu \in \mathscr{G}$, we have $\lambda^*\mu^* = (\lambda\mu)^*$. Moreover λ^* maps Z onto Z so we can extend this map in a natural way to a ring automorphism of E.

We observe now that $\mathscr{G}$ has no proper subgroups of finite index. First H/Δ^+ is a finitely generated torsion free abelian group so say $H/\Delta^+ = W_1 \times W_2 \times \cdots \times W_n$, a direct product of n infinite cyclic groups. Then clearly $\mathscr{G} = \mathrm{Hom}(H/\Delta^+, K^*) \simeq K^* \times K^* \times \cdots \times K^*$, a direct product of n copies of K^*. If $\mathscr{H}$ is a subgroup of $\mathscr{G}$ of finite index m, then $\mathscr{H}$ contains all mth powers of elements of $\mathscr{G}$. However K is algebraically closed so every element of K^* is an mth power and therefore $\mathscr{H} = \mathscr{G}$.

(i) Since E is a finite dimensional F-algebra, it follows that E has only finitely many central idempotents. Moreover since any central idempotent of $K[H] \subseteq E$ is clearly central in E, we conclude that $K[H]$ has only finitely many central idempotents also. Now $\mathscr{G}$ acts on $K[H]$ as ring automorphisms and hence $\mathscr{G}$ permutes the finitely many central idempotents. But $\mathscr{G}$ has no proper subgroups of finite index, so $\mathscr{G}$ must fix all such central idempotents and in particular $\mathscr{G}$ fixes e.

Write $e = \sum a_x \cdot x$. If $\lambda \in \mathscr{G}$, then

$$\sum a_x \cdot x = e = \lambda^*(e) = \sum a_x \lambda(x) \cdot x$$

and therefore for all $x \in \mathrm{Supp}\, e$ we have $\lambda(x) = 1$. Since H/Δ^+ is K-complete by Lemma 17.1, this yields $\mathrm{Supp}\, e \subseteq \Delta^+$ and (i) is proved.

(ii) Suppose now that $K[H]$ is semiprime and let L be an ideal of E with square zero. Then $L \cap K[H]$ is an ideal of $K[H]$ with square zero and hence $L \cap K[H] = 0$ since $K[H]$ is semiprime. Now if $\eta^{-1}\alpha \in L$, then $\alpha = \eta(\eta^{-1}\alpha) \in L \cap K[H]$ so $\alpha = 0$. Thus $L = 0$ and E is semiprime. Since E is a finite dimensional algebra, this therefore implies that E is a semisimple Artinian ring. Hence E has only a finite number of two sided ideals.

Now $\mathscr{G}$ acts on E as a group of ring automorphisms and hence $\mathscr{G}$ permutes the finitely many ideals of E. But $\mathscr{G}$ has no proper subgroups of finite index, so $\mathscr{G}$ must fix all such ideals. In particular $\mathscr{G}$ fixes setwise the ideal $Z^{-1}I \subseteq E$ and hence $\mathscr{G}$ fixes setwise $\tilde{I} = K[H] \cap Z^{-1}I$. Clearly $\tilde{I}$ is an ideal in $K[H]$, $\tilde{I} \supseteq I$, and it is easy to see that

$$\tilde{I} = \{\alpha \in K[H] \mid \eta\alpha \in I \quad \text{for some} \quad \eta \in Z, \quad \eta \neq 0\}.$$

Let $\alpha \in \tilde{I}$ and write $\alpha = \sum \alpha_i x_i$ where $\alpha_i \in K[\Delta^+]$, $x_1 = 1$, and the x_i are in distinct cosets of Δ^+ in H. Since H/Δ^+ is K-complete by Lemma 17.1, we conclude from Lemma 17.3 that

$$\theta^+(\alpha) = \alpha_1 = \sum k_i \lambda_i^*(\alpha)$$

for some $\lambda_i \in \mathscr{G}$, $k_i \in K$. Now $\mathscr{G}$ leaves $\tilde{I}$ setwise invariant so $\lambda_i^*(\alpha) \in \tilde{I}$ for all i and hence $\theta^+(\alpha) \in \tilde{I}$. We have therefore shown that $\theta^+(\tilde{I}) \subseteq \tilde{I}$.

Now Δ^+ is finite so certainly $\theta^+(I)$ is finite dimensional over K; choose $\beta_1, \beta_2, \ldots, \beta_s \in I$ so that $\theta^+(\beta_1), \theta^+(\beta_2), \ldots, \theta^+(\beta_s)$ span $\theta^+(I)$ over K. By the above $\theta^+(\beta_i) \subseteq \theta^+(\tilde{I}) \subseteq \tilde{I}$ so there exists $\zeta_i \in Z$, $\zeta_i \neq 0$ with $\zeta_i \theta^+(\beta_i) \in I$. If $\zeta = \zeta_1 \zeta_2 \cdots \zeta_s$, then $\zeta\theta^+(\beta_i) \in I$ for all i so $\zeta\theta^+(I) \subseteq I$. Since ζ is central in $K[H]$, $\zeta \neq 0$, and since ζ is not a zero divisor in $K[H]$, the result follows.

We now come to the main result of this section.

Theorem 23.4. Let e be a central idempotent in the group ring $K[G]$. Then $\langle \text{Supp}\, e \rangle$, the subgroup of G generated by the support of e, is a finite normal subgroup of G. Moreover if K has characteristic $p > 0$, then Supp e consists of p'-elements.

Proof. Let F be the algebraic closure of K. Then e is a central idempotent in $F[G] \supseteq K[G]$ with the same support. Thus we may assume that K is algebraically closed.

Let $H = \langle \text{Supp}\, e \rangle$ so that H is a finitely generated subgroup of G. Since e is central, we have clearly $H \subseteq \Delta(G)$ and H is normal in G. Now e is a central idempotent in $K[H]$ and $H = \Delta(H)$ so Lemma 23.3(i) yields Supp $e \subseteq \Delta^+(H)$. Thus $H = \Delta^+(H)$ is finite and the result follows from Theorem 23.2.

The following example of [*49*] shows that the assumption that e is central in the above is necessary in order to conclude that $\langle \text{Supp}\, e \rangle$ is finite. Let K be algebraically closed and let H be a finite nonabelian group whose order is prime to the characteristic of K. Then $K[H]$ is semisimple and not commutative and hence by the Wedderburn theorems $K[H]$ contains as a subring K_2, the ring of 2×2 matrices over K. Let $\mathscr{E}$, $\eta \in K_2 \subseteq K[H]$ be the elements

$$\mathscr{E} = \begin{pmatrix} 1 & 0 \\ 0 & 0 \end{pmatrix}, \qquad \eta = \begin{pmatrix} 0 & 1 \\ 0 & 0 \end{pmatrix}$$

so that $\mathscr{E}^2 = \mathscr{E}$, $\eta^2 = 0$, $\mathscr{E}\eta = \eta$, $\eta\mathscr{E} = 0$. Let $W = \langle x \rangle$ be infinite cyclic and set $G = H \times W$. Then $e = \mathscr{E} + \eta x \in K[G]$ and since x commutes with $\mathscr{E}$, η we have $e^2 = e$. Moreover for some $h \in H$, $hx \in \operatorname{Supp} e$ so $\langle \operatorname{Supp} e \rangle$ is infinite.

Now let E be an algebra over a field K. An element $\alpha \in E$ is said to be algebraic if $f(\alpha) = 0$ for some nonzero polynomial $f(\zeta) \in K[\zeta]$. If furthermore f is separable, that is, if f has distinct roots in the algebraic closure of K, then α is separably algebraic.

Corollary 23.5. Let α be a central element of the group ring $K[G]$ which is separably algebraic over K. Then $\langle \operatorname{Supp} \alpha \rangle$ is a finite normal subgroup of G. Moreover if K has characteristic $p > 0$, then $\operatorname{Supp} \alpha$ consists of p'-elements.

Proof. If F is the algebraic closure of K, then α is a central element of $F[G] \supseteq K[G]$ with the same support. Thus we may assume that K is algebraically closed.

Since α is separably algebraic over K, the algebra $K[\alpha]$ is semisimple and hence α is a K-linear sum of idempotents $e_1, e_2, \ldots, e_n \in K[\alpha]$. Moreover each e_i is central in $K[G]$. Finally

$$\operatorname{Supp} \alpha \subseteq \bigcup_{1}^{n} \operatorname{Supp} e_i$$

so the result follows from Theorem 23.4.

§24. REGULAR RINGS

A ring R is said to be regular in the sense of Von Neumann if for any $\alpha \in R$ there exists $\alpha' \in R$ with

$$\alpha\alpha'\alpha = \alpha.$$

In this section we obtain necessary and sufficient conditions for $K[G]$ to be regular. The following characterization of regular rings is due to Von Neumann in [*56*].

Lemma 24.1. A ring R is regular if and only if every finitely generated right ideal is generated by an idempotent. In particular, if R is Artinian and semisimple, then R is regular.

Proof. Suppose first that R satisfies the above condition on ideals and let $\alpha \in R$. Then αR is a finitely generated right ideal and hence there exists an idempotent $e \in R$ with $\alpha R = eR$. Since $e \in \alpha R$ we have $e = \alpha\alpha'$

for some $\alpha' \in R$ and since $\alpha \in eR$ we have

$$\alpha = e\alpha = \alpha\alpha'\alpha$$

so R is regular.

Now suppose that R is regular and let αR be a right ideal in R. Then there exists $\alpha' \in R$ with $\alpha\alpha'\alpha = \alpha$. Set $e = \alpha\alpha'$. Then

$$e^2 = (\alpha\alpha'\alpha)\alpha' = \alpha\alpha' = e$$

so e is an idempotent and $e \in \alpha R$. On the other hand, $e\alpha = \alpha\alpha'\alpha = \alpha$ so $\alpha \in eR$. Thus $\alpha R = eR$ and we have shown that every singly generated right ideal of R is generated by an idempotent.

Now consider the ideal $\alpha R + \beta R$. By the above $\alpha R = eR$ for some idempotent e. Since $\beta R \subseteq e\beta R + (1 - e)\beta R$, it follows easily that

$$\alpha R + \beta R = eR + (1 - e)\beta R.$$

Let f be an idempotent which generates $(1 - e)\beta R$. Then certainly $f^2 = f$ and $ef = 0$. Set $g = f(1 - e)$. Then

$$gf = f(1 - e)f = f^2 = f$$
$$g^2 = gf(1 - e) = f(1 - e) = g$$
$$eg = 0 = ge.$$

Now $g \in fR$ and $f = gf \in gR$, so $fR = gR$ and

$$\alpha R + \beta R = eR + gR.$$

Finally since e and g are orthogonal idempotents, $e + g$ is an idempotent, and $eR + gR = (e + g)R$. The result now follows easily by induction on the finite number of generators of the given right ideal.

Let R be an arbitrary ring and let I be any nonempty subset of R. We define the left annihilator $\ell(I)$ of I by

$$\ell(I) = \{\alpha \in R \mid \alpha I = 0\}.$$

It follows easily that $\ell(I)$ is a left ideal in R. In a similar manner, we can define $r(I)$, the right annihilator of I. Note that if $I = eR$ is a right ideal generated by an idempotent $e \neq 1$, then $1 - e \in \ell(I)$ and $1 - e \neq 0$. Thus Lemma 24.1 yields immediately

Lemma 24.2. Let R be a regular ring and let I be a finitely generated right ideal of R with $I \neq R$. Then $\ell(I) \neq 0$.

We now study group rings.

Lemma 24.3. Let $x_1, x_2, \ldots, x_n$ be a finite number of elements of G and let $H = \langle x_1, x_2, \ldots, x_n\rangle$ be the subgroup of G they generate. If I is the finitely generated right ideal of $K[G]$ given by

$$I = (1 - x_1)K[G] + (1 - x_2)K[G] + \cdots + (1 - x_n)K[G]$$

then $\ell(I) = 0$ if H is infinite and $\ell(I) = K[G]\omega(H)$ if H is finite where $\omega(H) = \sum_{x\in H} x$.

Proof. Let $\alpha \in \ell(I)$ so that for all i, $\alpha(1 - x_i) = 0$ and hence $\alpha x_i = \alpha$. This implies easily that for all $x \in H = \langle x_1, x_2, \ldots, x_n\rangle$ we have $\alpha x = \alpha$ and thus the coefficients of α are constant on the right cosets of H. Since Supp α is finite, this implies that $\alpha = 0$ if H is infinite and $\alpha \in K[G]\omega(H)$ if H is finite. Finally $x_i \in H$ implies that $\omega(H)\cdot(1 - x_i) = 0$ so the result follows.

Theorem 24.4 (Villamayor [*55*], Connell [*11*]). The group ring $K[G]$ is regular if and only if G is locally finite and has no elements of order p in case K has characteristic $p > 0$.

Proof. Suppose first that G is locally finite and has no elements of order p in case K has characteristic $p > 0$. Let $\alpha \in K[G]$. Since Supp α is finite, there exists a finitely generated and hence finite subgroup H of G with $\alpha \in K[H]$. Moreover $|H| \neq 0$ in K so by Theorem 15.3, $K[H]$ is Artinian and semisimple. Thus $K[H]$ is regular by Lemma 24.1 so there exists $\alpha' \in K[H] \subseteq K[G]$ with $\alpha\alpha'\alpha = \alpha$. Since α was arbitrary, $K[G]$ is regular.

Now suppose that $K[G]$ is regular and let ρ denote the natural epimorphism $\rho: K[G] \to K[G/G] = K$. Let $\{x_1, x_2, \ldots, x_n\}$ be a finite subset of G and set $H = \langle x_1, x_2, \ldots, x_n\rangle$. If I is the finitely generated right ideal

$$I = (1 - x_1)K[G] + (1 - x_2)K[G] + \cdots + (1 - x_n)K[G]$$

then I is contained in the kernel of ρ so $I \neq K[G]$. By Lemma 24.2, $\ell(I) \neq 0$ and hence by Lemma 24.3, H is finite. Thus G is locally finite.

Let $x \in G$ so that $\alpha = 1 - x \in K[G]$. By regularity there exists $\alpha' \in K[G]$ with $\alpha\alpha'\alpha = \alpha$ or

$$(1 - (1 - x)\alpha')\cdot(1 - x) = 0.$$

Thus

$$1 - (1 - x)\alpha' \in \ell((1 - x)K[G]) = K[G]\omega(\langle x\rangle)$$

by Lemma 22.3. Therefore if x has order m, then for some $\beta \in K[G]$ we have

$$1 - (1 - x)\alpha' = \beta(1 + x + x^2 + \cdots + x^{m-1})$$

and applying the homomorphism ρ to this expression yields $1 = \rho(\beta) \cdot m$. Hence $m \neq 0$ in K and the result follows.

§25. ANNIHILATOR IDEALS

An annihilator ideal of a ring R is a two sided ideal of R which is the left annihilator of some subset of R. Such ideals arise as follows.

Lemma 25.1. Let A be a subset of $K[G]$. Then A is an annihilator ideal if and only if $A = \ell(B)$ for some ideal B.

Proof. Suppose first that $A = \ell(B)$. Since A is a left annihilator, it is clearly a left ideal. Therefore it suffices to show that $Ax \subseteq A$ for all $x \in G$. Now B is an ideal and x is invertible so $B = x^{-1}B$. Hence

$$(Ax)B = (Ax)(x^{-1}B) = AB = 0$$

and $Ax \subseteq \ell(B) = A$.

Conversely suppose that A is an annihilator ideal and say $A = \ell(S)$ where S is a nonempty subset of $K[G]$. Let $B = K[G] \cdot S \cdot K[G]$ be the two sided ideal generated by S. Since $B \supseteq S$, we have $\ell(B) \subseteq \ell(S) = A$. On the other hand, since A is an ideal we have

$$AB = AK[G] \cdot S \cdot K[G] \subseteq AS \cdot K[G] = 0$$

so $A \subseteq \ell(B)$ and the result follows.

In this section we study annihilator ideals in $K[G]$ and we show that if $K[G]$ is semiprime, then such ideals are generated by their intersection with $K[\Delta^+(G)]$.

Theorem 25.2 (Passman [*40*], Smith [*50*]). Let $K[G]$ be a semiprime group ring and let $A_1, A_2, \ldots, A_n$ be ideals in $K[G]$ with $A_1A_2 \cdots A_n = 0$. Then the sets $\theta^+(A_i)$ are ideals in $K[\Delta^+(G)]$ and

$$\theta^+(A_1)\theta^+(A_2) \cdots \theta^+(A_n) = 0.$$

Proof. Since $\theta^+(\alpha) + \theta^+(\beta) = \theta^+(\alpha + \beta)$, $\theta^+(A_i)$ is clearly closed under addition. Furthermore, if $\alpha \in A_i$ and $\gamma \in K[\Delta^+]$, then $\alpha\gamma \in A_i$, $\gamma\alpha \in A_i$, and we have easily $\theta^+(\alpha\gamma) = \theta^+(\alpha)\gamma$, $\theta^+(\gamma\alpha) = \gamma\theta^+(\alpha)$. Thus $\theta^+(A_i)$ is an ideal in $K[\Delta^+]$.

By Lemma 20.1(ii), the sets $B_i = \theta(A_i)$ are ideals in $K[\Delta]$ and $B_1B_2 \cdots B_n = 0$. Moreover since $\Delta^+ \subseteq \Delta$, we have clearly

$$\theta^+(A_i) = \theta^+(\theta(A_i)) = \theta^+(B_i).$$

Choose $\beta_i \in B_i$ for $i = 1, 2, \ldots, n$ and let H be the subgroup of G generated by the supports of this finite number of elements. Then H is a finitely generated subgroup of $\Delta(G)$ and $\beta_i \in I_i = B_i \cap K[H]$.

Let F denote the algebraic closure of the field K. By assumption $K[G]$ is semiprime so by Theorem 3.7, $\Delta(G)$ has no elements of order p in case K has characteristic p. Hence by Theorems 3.3 and 3.7, $F[H]$ is semiprime. Since $F \cdot I_i$ is an ideal in $F[H]$, Lemma 23.3(ii) applies and there exist elements $\zeta_i \neq 0$ which are central in $F[H]$ and which are not zero divisors in $F[H]$ such that $\zeta_i \theta^+(F \cdot I_i) \subseteq F \cdot I_i$. Observe that since $\Delta^+(H)$ is the set of torsion elements of H, we have $\Delta^+(H) = H \cap \Delta^+(G)$ and hence the map $\theta^+ : F[H] \to F[\Delta^+(H)]$ is just the restriction to $F[H]$ of the map $\theta^+ : F[G] \to F[\Delta^+(G)]$. Thus

$$\zeta_1\theta^+(\beta_1) \cdot \zeta_2\theta^+(\beta_2) \cdots \zeta_n\theta^+(\beta_n) \in F \cdot I_1 I_2 \cdots I_n = 0$$

and hence $\theta^+(\beta_1)\theta^+(\beta_2) \cdots \theta^+(\beta_n) = 0$ by the properties of the ζ_i. We have therefore shown that $\theta^+(B_1)\theta^+(B_2) \cdots \theta^+(B_n) = 0$ and since $\theta^+(A_i) = \theta^+(B_i)$ the result follows.

Theorem 25.3 (Smith [*50*]). Let $K[G]$ be a semiprime group ring and let A be an annihilator ideal in $K[G]$. Then

$$A = \theta^+(A) \cdot K[G] = (A \cap K[\Delta^+]) \cdot K[G].$$

Proof. We remark first that for any ideal I in $K[G]$ we have clearly $I \subseteq \theta^+(I) \cdot K[G]$. Hence $A \subseteq \theta^+(A) \cdot K[G]$.

Now by Lemma 25.1, $A = \ell(B)$ where B is an ideal in $K[G]$. Then $AB = 0$ so by Theorem 25.4 we have $\theta^+(A)\theta^+(B) = 0$. Since $B \subseteq \theta^+(B) \cdot K[G]$ this yields $\theta^+(A) \subseteq \ell(B) = A$ and thus $\theta^+(A) \cdot K[G] \subseteq A$. Therefore $A = \theta^+(A) \cdot K[G]$. Since this implies that $A \cap K[\Delta^+] \supseteq \theta^+(A)$, the result follows.

The preceding two results are decidedly false without the semiprime assumption as the following example shows. Let G be the abelian group $G = \langle x \rangle \times \langle a \rangle \times \langle b \rangle$ where x has infinite order and a and b have prime order p. Then $G = \Delta(G)$ and $\Delta^+(G) = \langle a \rangle \times \langle b \rangle$. Let K be a field of characteristic p and let $\alpha, \beta \in K[G]$ be given by

$$\alpha = (1 - a) + x(1 - b)$$
$$\beta = (1 - a)^{p-2}(1 - b)^{p-2}\big((1 - b) - x^{-1}(1 - a)\big).$$

Since $(1 - a)^p = 1 - a^p = 0$ and $(1 - b)^p = 0$ we have easily $\alpha\beta = 0$. Set $B = \beta K[G]$ so that B is an ideal in the commutative ring $K[G]$.

Thus if $A = \ell(B)$, then A is an annihilator ideal and clearly $\alpha \in A$. Now $\theta^+(A)$ contains $\theta^+(\alpha) = 1 - a$, $\theta^+(B)$ contains

$$\theta^+(\beta) = (1-a)^{p-2}(1-b)^{p-1},$$

and

$$\theta^+(\alpha)\theta^+(\beta) = (1-a)^{p-1}(1-b)^{p-1} \neq 0$$

since $\langle a \rangle \times \langle b \rangle$ is a direct product. Thus $\theta^+(A)\theta^+(B) \neq 0$ even though $AB = 0$. Moreover $\theta^+(\alpha)\beta \neq 0$ so $\theta^+(A) \nsubseteq \ell(B) = A$.

Certainly every commutative ring has a complete ring of quotients and in particular this applies to the center of $K[G]$.

Theorem 25.4 (Smith [*50*]). Let $K[G]$ be a semiprime group ring.

(i) Let A be an annihilator ideal in $K[G]$ and let $\alpha \in A$, $\alpha \neq 0$. Then there exists a central idempotent $e \in A$ with $e\alpha = \alpha$.

(ii) The complete ring of quotients of the center of $K[G]$ is a regular ring.

Proof. (i) Write $\alpha = \sum \alpha_i x_i$ where $\alpha_i \in K[\Delta^+]$ and the x_i are in distinct cosets of Δ^+ in G. Then Theorem 25.3 implies that $\alpha_i \in A$ for all i. Let H be the normal subgroup of G generated by the elements in Supp α_i and their finitely many conjugates for all i. Then H is a finitely generated subgroup of Δ^+ and hence H is clearly finite by Lemma 2.2. Moreover we have $\alpha_i \in I = A \cap K[H]$ for all i.

Since $K[G]$ is semiprime, Theorem 3.7 implies that H has no elements of order p in case K has characteristic p and thus $K[H]$ is a semisimple finite dimensional algebra by Theorems 3.2 and 3.7. Therefore by the Wedderburn theorems, since $I \neq 0$, I has an identity element $e \neq 0$. Then e is an idempotent and $e\alpha_i = \alpha_i$ so $e\alpha = \alpha$. Now if $x \in G$, then conjugation by x fixes A and $K[H]$ setwise so x acts as an algebra automorphism on I. Therefore x fixes the unique identity element e. Since $K[G]$ is spanned by the elements of G, this shows that e is central in $K[G]$ and (i) follows.

(ii) Let C denote the center of $K[G]$ and let Q be its complete ring of quotients. Let $\alpha = \beta^{-1}\gamma \in Q$ with $\beta, \gamma \in C$ and with β not a zero divisor in C. If $\alpha = 0$, then $\alpha\alpha'\alpha = \alpha$ with $\alpha' = 0$. Thus let $\alpha \neq 0$ so that $\gamma \neq 0$.

Let $\ell(S)$ denote the left annihilator of the set S in $K[G]$. Since $\gamma \in C$, it follows that $\ell(\gamma)$ is an ideal in $K[G]$ and thus $A = \ell(\ell(\gamma))$ is an annihilator ideal by Lemma 25.1. Moreover clearly $\gamma \in A$ and thus by (i) above there exists an idempotent $e \in C$ such that $e \in A$ and $e\gamma = \gamma$.

Set $\delta = \gamma + (1 - e)$ so that $\delta \in C$. We show now that δ is not a zero divisor in C or in fact in $K[G]$. Suppose $\delta\varepsilon = 0$. Then

$$0 = (1 - e)\delta \cdot \varepsilon = (1 - e)\varepsilon$$
$$0 = e\delta \cdot \varepsilon = \gamma\varepsilon.$$

The latter implies that $\varepsilon \in \ell(\gamma)$ so since $e \in \ell(\ell(\gamma))$ we have $e\varepsilon = 0$. Thus $(1 - e)\varepsilon = 0$ and $e\varepsilon = 0$ so $\varepsilon = 0$ and $\delta^{-1} \in Q$. Now $e\delta = e\gamma = \gamma$ so $e = \delta^{-1}\gamma$ and if we set $\alpha' = \delta^{-1}\beta \in Q$, then since C is commutative we have

$$\begin{aligned}\alpha\alpha'\alpha &= \beta^{-1}(\delta^{-1}\gamma)\gamma = \beta^{-1}e\gamma \\ &= \beta^{-1}\gamma = \alpha\end{aligned}$$

and Q is regular.

It was observed in [*50*] that many results on group rings apply equally well to group rings modulo an annihilator ideal. The following is a typical example which is related to the arguments of Section 7. Let R be a ring with center Z. We say that R has dimension less than m over Z if given any m elements $\alpha_1, \alpha_2, \ldots, \alpha_m \in R$, there exists $z_1, z_2, \ldots, z_m \in Z$ not all zero with $z_1\alpha_1 + z_2\alpha_2 + \cdots + z_m\alpha_m = 0$. This of course reduces to the usual definition in case Z is a field.

Theorem 25.5 (Smith [*50*]). Let I be an annihilator ideal in $K[G]$ and suppose that $K[G]/I$ is an order in a ring Q. If Q has dimension less than m over its center Z, then $[G:\Delta] < m$.

Proof. If $\gamma \in K[G]$, we let $\bar{\gamma}$ denote its image in $K[G]/I \subseteq Q$. Suppose by way of contradiction that $[G:\Delta] \geq m$ and let $g_1, g_2, \ldots, g_m \in G$ be elements in distinct cosets of Δ. By definition there exists $z_1, z_2, \ldots, z_m \in Z$ not all zero with

$$\bar{g}_1z_1 + \bar{g}_2z_2 + \cdots + \bar{g}_mz_m = 0.$$

Now $K[G]/I$ is an order in Q so there exists $\alpha_i, \beta_i \in K[G]$ with $z_i = \bar{\alpha}_i^{-1}\bar{\beta}_i$. Set $\alpha = \alpha_1\alpha_2 \cdots \alpha_m \in K[G]$. Then $\bar{\alpha}$ is invertible in Q and since z_i is central

$$\begin{aligned}\bar{\alpha}z_i &= \bar{\alpha}_1 \cdots \bar{\alpha}_{i-1}(\bar{\alpha}_iz_i)\bar{\alpha}_{i+1} \cdots \bar{\alpha}_m \\ &= \bar{\gamma}_i \in K[G]/I\end{aligned}$$

for some $\gamma_i \in K[G]$. Thus $z_i = \bar{\alpha}^{-1}\bar{\gamma}_i$.

Let $x \in G$. Since the z_i are central in Q, we have

$$\sum_1^m \bar{g}_i\bar{x}\bar{\gamma}_i = \left(\sum_1^m \bar{g}_iz_i\right)\bar{x}\bar{\alpha} = 0$$

and thus $\sum_1^m g_i x\gamma_i \in I$. Now $I = \ell(T)$ for some subset $T \subseteq K[G]$ and let $\tau \in T$. Then $I\tau = 0$ yields $\sum_1^m g_i x\gamma_i\tau = 0$.

Fix a subscript j. Then for all $x \in G$

$$\sum_{i=1}^{m} (g_j^{-1}g_i)x(\gamma_i\tau) = 0$$

so by Lemma 5.3 there exists $y \in G$ with

$$\sum_{i=1}^{m} \theta(g_j^{-1}g_i)^y\gamma_i\tau = 0.$$

Now for $i \neq j$, $g_j^{-1}g_i \notin \Delta$ so $\theta(g_j^{-1}g_i) = 0$ and for $i = j$, $\theta(g_j^{-1}g_i) = 1$. Thus we obtain $\gamma_j\tau = 0$ and since this holds for all $\tau \in T$ we have $\gamma_j \in \ell(T) = I$. Thus $\bar{\gamma}_j = 0$ so $z_j = 0$ for all j, a contradiction, and the result follows.

§26. ZERO DIVISORS

The question of the existence of proper zero divisors in group rings is probably the oldest and least understood problem in the field. In essence all we do here is define a property of a group which obviously implies that $K[G]$ has no zero divisors and then try to show that groups with this property form a fairly large class.

An element $\alpha \in K[G]$ is a unit if α is invertible, that is, if there exists $\beta \in K[G]$ with $\alpha\beta = \beta\alpha = 1$. Of course all elements of the form $\alpha = kx$ with $k \in K$, $k \neq 0$, $x \in G$ are units and we consider these to be trivial. In this section we also study the question of the existence of nontrivial units in group rings.

Lemma 26.1. Let G be a group which is not torsion free. Then $K[G]$ has proper divisors of zero. Moreover if $|K| > 3$, then $K[G]$ has nontrivial units.

Proof. By assumption G contains a finite subgroup $H \neq \langle 1 \rangle$. Set

$$\alpha = \sum_{x \in H} x \in K[G].$$

If $h \in H$, then $hH = H$ so $h\alpha = \alpha$. This yields easily $\alpha^2 = n\alpha$ where $n = |H|$. From $\alpha(\alpha - n) = 0$, we see that $K[G]$ has proper divisors of zero. Now suppose that $|K| > 3$ and choose $a \in K$ with $a \neq 0$, 1 and $a \cdot n \neq 1$. Then $1 - a\alpha$ is a nontrivial unit in $K[G]$ with inverse $1 - b\alpha$ where $b = a/(an - 1)$.

The assumption on $|K|$ above is needed. For example, if $|G| = 2$ and if $|K| = 2$ or 3, then all units in $K[G]$ are trivial. The question of interest here is whether the converse of the above lemma is true. Namely if G is torsion free, does it follow that $K[G]$ has no zero divisors and only trivial units. The answer is not known and we offer only special cases here.

A group G is said to be a u.p.-group (unique product group) if given any two nonempty finite subsets A and B of G, then there exists at least one element $x \in G$ which has a unique representation in the form $x = ab$ with $a \in A$ and $b \in B$.

A group G is said to be a t.u.p.-group (two unique products group) if given any two nonempty finite subsets A and B of G with $|A| + |B| > 2$, then there exists at least two distinct elements x and y of G which have unique representations in the form $x = ab$, $y = cd$ with $a, c \in A$ and $b, d \in B$.

It is trivial to see that every t.u.p.-group is a u.p.-group.

Theorem 26.2. Let K be an arbitrary field. If G is a u.p.-group, then $K[G]$ has no proper divisors of zero. If G is a t.u.p.-group, then $K[G]$ has only trivial units.

Proof. Let α and β be nonzero elements of $K[G]$ and set $A = \operatorname{Supp} \alpha$, $B = \operatorname{Supp} \beta$. If G is a u.p.-group and if $x \in G$ is a uniquely represented element of AB, then it follows easily that $x \in \operatorname{Supp} \alpha\beta$ and hence $\alpha\beta \neq 0$. Thus $K[G]$ has no proper divisors of zero. Suppose in addition that α is not a trivial unit so $|A| > 1$. If G is a t.u.p.-group and if $x, y \in G$ are two uniquely represented elements of AB, then $x, y \in \operatorname{Supp} \alpha\beta$ so certainly $\alpha\beta \neq 1$. Thus $K[G]$ has no nontrivial units.

In view of Lemma 26.1 this implies that every u.p.-group is torsion free. The following result shows that such groups form a fairly large class.

Theorem 26.3 (G. Higman [*18*], Rudin–Schneider [*44*]). Let G be a group. Then any of the following implies that G is a u.p.-group (respectively, a t.u.p.-group).

(i) G has a normal subgroup H such that both H and G/H are u.p.-groups (respectively, t.u.p.-groups).

(ii) G has a family of normal subgroups H_ν such that $\cap_\nu H_\nu = \langle 1 \rangle$ and such that G/H_ν is a u.p.-group (respectively, a t.u.p.-group).

(iii) Every finitely generated nonidentity subgroup of G can be mapped homomorphically onto a nonidentity u.p.-group (respectively, a nonidentity t.u.p.-group).

Proof. The proofs for the two types of groups are essentially the same. Therefore we will only consider the slightly harder case of t.u.p.-groups.

Let A and B be finite nonempty subsets of G with $|A| + |B| > 2$. We prove that there exist two uniquely represented elements in the product AB by induction on $|A| + |B|$. The case $|A| + |B| = 3$ is obvious. Choose $g, h \in G$ so that $1 \in gA$, $1 \in Bh$. Now every product of elements of the sets gA and Bh is of the form $gabh$. Since $gabh = ga'b'h$ if and only if $ab = a'b'$, it clearly suffices to consider instead the sets gA and Bh or in other words we may assume that $1 \in A$ and $1 \in B$.

Let us suppose that G has subgroups L and N such that $A, B \subseteq L$, N is normal in L, and L/N is a t.u.p.-group. Moreover if $\bar{A}$ and $\bar{B}$ are the images of A and B in L/N, suppose that $|\bar{A}| + |\bar{B}| > 2$. Since L/N is a t.u.p.-group, there are nonempty subsets A_1, A_2 of A and B_1, B_2 of B such that for any $a \in A$, $b \in B$ then $ab \in A_iB_i$ implies that $a \in A_i$, $b \in B_i$ for $i = 1, 2$. Moreover $|A_i| + |B_i| < |A| + |B|$ and $A_1B_1 \cap A_2B_2 = \varnothing$. Thus by induction if $|A_i| + |B_i| > 2$ or clearly if $|A_i| + |B_i| = 2$, there exists uniquely represented elements $x \in A_1B_1$ and $y \in A_2B_2$. These are then uniquely represented elements in the product AB and this proves the induction step. We now consider each of the three parts of this theorem separately.

(i) Here we take $L = G$ and $N = H$. If $|\bar{A}| + |\bar{B}| > 2$, we are done by the above. If $|\bar{A}| + |\bar{B}| \leq 2$, then since $1 \in A$ and $1 \in B$ we have $A \subseteq H$ and $B \subseteq H$. Then AB has two unique products since H is a t.u.p.-group and the result follows here.

(ii) Now $1 \in A$, $1 \in B$, and $|A| + |B| > 2$. Thus there exists $z \in A \cup B$ with $z \neq 1$. By assumption, there exists a normal subgroup H_ν such that $z \notin H_\nu$ and such that G/H_ν is a t.u.p.-group. Thus if we take $L = G$ and $N = H_\nu$, then $|\bar{A}| + |\bar{B}| > 2$ and we are done by the above.

(iii) Here we let L be the subgroup of G generated by the elements of A and of B. Then L is a finitely generated nonidentity subgroup of G so by assumption there exists a normal subgroup N of L, $N \neq L$, such that L/N is a t.u.p.-group. Since $1 \in A$, $1 \in B$, $|\bar{A}| + |\bar{B}| \leq 2$ implies that A, $B \subseteq N$ so $L = N$, a contradiction. Thus $|\bar{A}| + |\bar{B}| > 2$ and the result follows.

The above theorem shows how to construct new t.u.p.-groups from old ones but it does not show that such groups exist. We do this below.

A group G is an ordered group if it admits a strict linear ordering $<$ such that $x < y$ implies $xz < yz$ and $zx < zy$ for all $z \in G$.

Lemma 26.4. Any ordered group is a t.u.p.-group.

Proof. Suppose G is an ordered group and A and B are finite nonempty subsets of G with $|A| + |B| > 2$. Let a^+ and b^+ be the maximal elements of A and B, respectively, and let a^- and b^- be their minimal elements. Then clearly $x = a^+b^+$ and $y = a^-b^-$ are distinct uniquely represented elements of AB.

There are many group theoretic theorems which guarantee that certain large classes of groups can be ordered. We will content ourselves here with just one example.

Lemma 26.5. A group G is an ordered group if and only if G has a subset S such that

(i) $x, y \in S$ implies that $xy \in S$.

(ii) $x^{-1}Sx = S$ for all $x \in G$.

(iii) $1 \notin S$ and if $x \in G$, $x \neq 1$, then either x or x^{-1} belongs to S.

Proof. Suppose that G has an ordering $<$. Then it is easy to see that $S = \{x \in G \mid 1 < x\}$ has the required properties.

Conversely suppose that we are given S. We say $x < y$ if and only if $x^{-1}y \in S$. Then condition (i) implies that $<$ is transitive, (ii) implies that $<$ is compatible with group multiplication, and (iii) implies that $<$ is a strict linear ordering.

Lemma 26.6. Any torsion free abelian group is an ordered group.

Proof. Let G be a torsion free abelian group and let $\mathscr{S}$ be the family of all subsets S of G which satisfy the condition $x, y \in S$ implies $xy \in S$ and also $1 \notin S$. It follows easily by Zorn's lemma that $\mathscr{S}$ contains a maximal member S. We show that S satisfies (i), (ii), and (iii) of the previous lemma. Now (i) is given and (ii) follows since G is abelian. We consider (iii). Let $x \in G$, $x \neq 1$, and suppose by way of contradiction that neither x nor x^{-1} belongs to S. If

$$T = S \cup \{sx^n \mid s \in S, n \geq 1\} \cup \{x^n \mid n \geq 1\}$$

then clearly $T > S$ and T is closed under multiplication. By the maximality of S in $\mathscr{S}$ we have $T \notin \mathscr{S}$ so $1 \in T$. Now G is torsion free so $1 \notin \{x^n \mid n \geq 1\}$ and therefore for some $s \in S$ and $n \geq 1$, we have $1 = sx^n$ so $x^{-n} \in S$. Replacing x by x^{-1} in this argument we conclude also that $x^m \in S$ for some $m \geq 1$. Since S is closed under multiplication this yields

$$1 = (x^m)^n(x^{-n})^m \in S$$

a contradiction. Thus (iii) is satisfied and Lemma 26.5 yields the result.

Theorem 26.7 (Bovdi [9]). Let G be a group and suppose that $G = G_0 \supseteq G_1 \supseteq G_2 \supseteq \cdots \supseteq G_n = \langle 1 \rangle$ where G_{i+1} is normal in G_i and G_i/G_{i+1} is torsion free abelian. If K is any field, then $K[G]$ has no proper divisors of zero and no nontrivial units.

Proof. By Lemmas 26.4 and 26.6, G_i/G_{i+1} is a t.u.p.-group. Thus by Theorem 26.3(i) and induction, G is a t.u.p.-group. The result follows from Theorem 26.2.

We remark that not every t.u.p.-group is an ordered group. For example let $G = \langle x \rangle \times_\sigma \langle y \rangle$ be the semidirect product of two infinite cyclic groups with $x^y = x^{-1}$. Since both $\langle x \rangle$ and $G/\langle x \rangle \simeq \langle y \rangle$ are ordered groups by Lemma 26.6, it follows from Theorem 26.3(i) and Lemma 26.4 that G is a t.u.p.-group. On the other hand G is not an ordered group since $x > x^{-1}$ implies

$$x^{-1} = y^{-1}xy > y^{-1}x^{-1}y = x$$

and this in turn implies that $x > x^{-1}$, a contradiction.

§27. GRADED POLYNOMIAL RINGS

There is a beautiful technique due to Golod and Shafarevitch which yields many interesting examples in class field theory, ring theory, and group theory. While there is at present no really direct application to group rings, we will nevertheless include this result here in anticipation of such applications.

Let K be a field and let $T = K[\zeta_1, \zeta_2, \ldots, \zeta_d]$ be the polynomial ring over K in the d noncommuting variables $\zeta_1, \zeta_2, \ldots, \zeta_d$. A monomial in T is an element μ of the form $\mu = \zeta_{i_1}\zeta_{i_2} \cdots \zeta_{i_r}$. Certainly these are all linearly independent and form a basis for T over K. Moreover to each μ we can associate its degree, namely the number of ζ_i's occurring in the product so that, for example, the above μ has degree r. For each integer n let T_n be the K-subspace of T spanned by all monomials of T of degree n. We then have clearly the vector space direct sum

$$T = T_0 + T_1 + \cdots + T_n + \cdots$$

and dim $T_n = d^n$ since there are d^n distinct monomials of degree n. The elements of the various T_n's are called the homogeneous elements of T.

Lemma 27.1. Let A be an ideal in T generated by homogeneous elements. Then

$$A = A_0 + A_1 + \cdots + A_n + \cdots$$

where $A_i = A \cap T_i$.

Proof. Certainly $A \supseteq \sum A_i$ and we need only show the reverse inclusion. Let A be generated by the homogeneous elements $f_\nu \in T$. Then every element of A is a finite sum of the form $\gamma = \sum_{i,\nu} \alpha_{i\nu} f_\nu \beta_{i\nu}$ with $\alpha_{i\nu}$, $\beta_{i\nu} \in T$. Now each $\alpha_{i\nu}$ and $\beta_{i\nu}$ is a K-linear sum of monomials and thus γ is a K-linear sum of terms of the form $\mu f_\nu \eta$ where μ and η are monomials. Since f_ν is homogeneous so is $\mu f_\nu \eta$ and thus $\mu f_\nu \eta \in A \cap T_n$ for some n.

Lemma 27.2. Let A be an ideal in T generated by the elements $f_1, f_2, f_3, \ldots$. Suppose that each f_i is homogeneous of degree $n_i \geq 1$. Then for all $n \geq 1$ we have

$$b_n \geq d b_{n-1} - \sum_{n_i \leq n} b_{n-n_i}$$

where $b_n = \dim T_n/(A \cap T_n)$.

Proof. By Lemma 27.1, $A = A_0 + A_1 + \cdots + A_n + \cdots$ where $A_n = A \cap T_n$. Since A_n is a K-subspace of T_n we can choose a complementary subspace B_n. Thus T_n is the vector space direct sum $T_n = A_n + B_n$ and by definition $\dim B_n = b_n$.

We show now that for $n \geq 1$, we have

$$A_n \subseteq \sum_{j=1}^{d} A_{n-1}\zeta_j + \sum_{n_i \leq n} B_{n-n_i} f_i.$$

Certainly A is spanned by elements of the form $\mu f_i \eta$ where μ and η are monomials and each such element is homogeneous. Thus it suffices to show that if $\gamma = \mu f_i \eta \in A_n$, then γ is a member of the above right-hand side. Suppose first that $\eta \neq 1$. Then $\eta = \eta'\zeta_j$ for some j so $\gamma = \gamma'\zeta_j$ where $\gamma' = \mu f_i \eta' \in A_{n-1}$. Hence $\gamma \in A_{n-1}\zeta_j$. Now suppose that $\eta = 1$. Since $\gamma = \mu f_i$ has degree n and f_i has degree n_i, it follows that $n_i \leq n$ and

$$\mu \in T_{n-n_i} = A_{n-n_i} + B_{n-n_i}.$$

Write $\mu = \alpha + \beta$ where $\alpha \in A_{n-n_i}$, $\beta \in B_{n-n_i}$ and since $n_i \geq 1$ write $f_i = \sum_1^d \gamma_j \zeta_j$ where γ_j is homogeneous of degree $n_i - 1$. Then

$$\begin{aligned}\gamma &= \mu f_i = \alpha f_i + \beta f_i \\ &= \sum_1^d (\alpha\gamma_j)\zeta_j + \beta f_i \in \sum_{j=1}^d A_{n-1}\zeta_j + B_{n-n_i} f_i\end{aligned}$$

since $\alpha\gamma_j$ is a homogeneous element of A of degree $n - 1$.

Let $a_n = \dim A_n$. Then clearly $a_{n-1} = \dim A_{n-1}\zeta_j$ and $b_{n-n_i} = \dim B_{n-n_i}f_i$ so the above inclusion yields

$$a_n \leq da_{n-1} + \sum_{n_i \leq n} b_{n-n_i}.$$

Since $a_n + b_n = \dim T_n = d^n$ we therefore obtain

$$b_n \geq db_{n-1} - \sum_{n_i \leq n} b_{n-n_i}$$

and the result follows.

***Theorem** 27.3* (Golod–Shafarevitch [*14*]). Let $T = K[\zeta_1, \zeta_2, \ldots, \zeta_d]$ be the polynomial ring over K in d noncommuting variables and let A be an ideal in T generated by the polynomials $f_1, f_2, f_3, \ldots$. Suppose that each f_i is homogeneous of degree $n_i \geq 2$ and that there are at most $(d-1)^2/4$ f_i's of any given degree. Then T/A is infinite dimensional.

Proof. If $d = 1$, then $(d-1)^2/4 = 0$ so there are no f_i's. Hence $A = 0$ and $T/A = T$ is infinite dimensional. Thus we will assume below that $d > 1$.

We will use the notation of the previous two lemmas. In view of Lemma 27.1 we have $\dim T/A = \sum_{n=0}^{\infty} b_n$ and hence it suffices to show that the b_n's, which are of course nonnegative integers, are nonzero. Since all $n_i \geq 2$ we have clearly $A_0 = 0 = A_1$ and thus $b_0 = 1$, $b_1 = d$. We prove below by induction on $n \geq 1$ that

$$b_n \geq \frac{d-1}{2} \sum_{i=0}^{n-1} b_i$$

and since $d > 1$, this will clearly yield the result. The case $n = 1$ is clear.

Suppose that $n > 1$ and the above result holds at $n - 1$. Then

$$b_{n-1} \geq \frac{d-1}{2} \sum_{i=0}^{n-2} b_i.$$

Now by Lemma 27.2

$$b_n \geq db_{n-1} - \sum_{n_i \leq n} b_{n-n_i}.$$

Consider the right-hand sum $\sum_{n_i \leq n} b_{n-n_i}$. First all $n_i \geq 2$ so b_n and b_{n-1} do not occur as summands. Moreover the number of times b_j occurs as a summand for $j \leq n-2$ is clearly equal to the number of f_i of degree $n - j$ and this is at most $(d-1)^2/4$ by assumption. Thus

$$\sum_{n_i \leq n} b_{n-n_i} \leq \frac{(d-1)^2}{4} \sum_{i=0}^{n-2} b_i$$

so that

$$
\begin{aligned}
b_n &\geq db_{n-1} - \sum_{n_i \leq n} b_{n-n_i} \\
&\geq \frac{d-1}{2} b_{n-1} + \frac{d+1}{2} b_{n-1} - \frac{(d-1)^2}{4} \sum_{i=0}^{n-2} b_i \\
&\geq \frac{d-1}{2} b_{n-1} + \left(\frac{(d+1)(d-1)}{4} - \frac{(d-1)^2}{4}\right) \sum_{i=0}^{n-2} b_i \\
&= \frac{d-1}{2} \sum_{i=0}^{n-1} b_i
\end{aligned}
$$

and the induction step is proved. The result follows.

As an application we have

Theorem 27.4 (Golod [*13*]). Let $T = K[\zeta_1, \zeta_2, \ldots, \zeta_d]$ be the polynomial ring over K in d noncommuting variables and let M be the maximal ideal of T generated by all the ζ_i. If K is countable and if $d \geq 3$, then there exists an ideal A of T contained in M such that M/A is an infinite dimensional nil algebra.

Proof. Observe that $M = T_1 + T_2 + \cdots$ and that for $n \geq 1$, $M^n = T_n + T_{n+1} + \cdots$. Since K is countable so is M and we can write $M = \{\alpha_1, \alpha_2, \alpha_3, \ldots\}$. We now define an increasing sequence of integers s_i as follows. First $s_1 = 2$ and then if s_n is given we choose $s_{n+1} > s_n$ minimal with

$$\alpha_n^{s_n} \in T_{s_n} + T_{s_n+1} + T_{s_n+2} + \cdots + T_{s_{n+1}-1}.$$

In this way the powers $\alpha_n^{s_n}$ all fall in nonoverlapping segments of $T_2 + T_3 + \cdots$ and we can define homogeneous polynomials $f_i \in T_i$ for $i \geq 2$ by

$$\alpha_n^{s_n} = f_{s_n} + f_{s_n+1} + f_{s_n+2} + \cdots + f_{s_{n+1}-1}.$$

Let A be the ideal of T generated by the nonzero f_i's. By construction there is at most one such generator of each degree ≥ 2 and also $1 \leq (d-1)^2/4$ since $d \geq 3$. Hence, by Theorem 27.4, T/A is infinite dimensional and therefore so is M/A. Moreover M/A is nil since $\alpha_n^{s_n} \in A$. This completes the proof.

The following result of Golod settles in the negative the so-called Kurosh problem.

Corollary 27.5. There exists a finitely generated nil ring (without 1) which is not nilpotent.

Proof. Let K be the prime field $GF(p)$ for some prime p and let $d = 3$ in the above theorem. If $R = M/A$ then by Theorem 27.4, R is an infinite dimensional nil ring which is generated as a ring by the images of ζ_1, ζ_2, and ζ_3. Moreover R is not nilpotent since otherwise $T_n \subseteq A$ for all sufficiently large n and then M/A would certainly be finite dimensional, a contradiction.

We will consider additional consequences in the next section.

§28. ALGEBRAIC RINGS

Let E be an algebra over K. Then E is said to be algebraic if every element of E is algebraic over K.

Theorem 28.1 (Herstein [*16*]). If G is locally finite, then the group ring $K[G]$ is algebraic over K. Conversely if K has characteristic 0 and if $K[G]$ is algebraic, then G is locally finite.

Proof. Suppose first that G is locally finite and let $\alpha \in K[G]$. Then there exists a finite subgroup H of G such that $\alpha \in K[H]$. Since $K[H]$ is a finite dimensional K-algebra, it follows that the set $\{1, \alpha, \alpha^2, \alpha^3, \ldots\}$ must be linearly dependent thereby showing that α is algebraic over K. Hence $K[G]$ is algebraic.

Now suppose that $K[G]$ is algebraic and that K has characteristic 0 and let $\{x_1, x_2, \ldots, x_t\}$ be a finite subset of G. By adding inverses if necessary, we can assume that $\{x_1, x_2, \ldots, x_t\}$ generate $H = \langle x_1, x_2, \ldots, x_t\rangle$ as a semigroup. Set $\alpha = x_1 + x_2 + \cdots + x_t \in K[G]$. Then by assumption α is algebraic over K so we have

$$\alpha^{n+1} = a_0 + a_1\alpha + \cdots + a_n\alpha^n$$

for some integer $n \geq 0$ and some $a_i \in K$.

Let $w = x_{i_1}x_{i_2} \cdots x_{i_{n+1}}$ be a word in the x_i's of length $n + 1$. Since all coefficients in α^{n+1} are positive integers and since K has characteristic 0, it follows that $w \in \operatorname{Supp} \alpha^{n+1}$. Now $\alpha^{n+1} = \sum a_i\alpha^i$ so $w \in \operatorname{Supp} \alpha^i$ for some $i \leq n$. This therefore says that every word of length $n + 1$ is equal to some word of shorter length and thus clearly every element of H can be written as a word of length at most n in $x_1, x_2, \ldots, x_t$. Hence H is finite and G is locally finite.

The characteristic p problem is still open but we can handle one special case quite easily. An algebra E is said to be algebraic of bounded degree if there exists an integer n such that all $\alpha \in E$ satisfy a polynomial of degree n over K.

Lemma 28.2. Let E be an algebraic algebra over K. Then JE is a nil ideal and $\alpha\beta = 1$ implies $\beta\alpha = 1$ for all $\alpha, \beta \in E$. Moreover if E is algebraic of bounded degree, then E satisfies a polynomial identity over K.

Proof. Let $\alpha \in E$. Since α is algebraic over K we have

$$(1 + a_1\alpha + \cdots + a_r\alpha^r)\alpha^s = 0$$

for suitable integers $r \geq 0$, $s \geq 0$, and $a_i \in K$. If $\alpha \in JE$, then $\gamma = a_1\alpha + \cdots + a_r\alpha^r \in JE$ so $1 + \gamma$ is invertible. Thus $\alpha^s = 0$ and JE is nil. If $\alpha\beta = 1$, then multiplying the above equation on the right by β^{s+1} yields

$$\beta = -(a_1 + a_2\alpha + \cdots + a_r\alpha^{r-1})$$

so α and β commute.

Now suppose that E is algebraic of bounded degree n and let $\alpha, \beta \in E$. Then by assumption

$$a_0 + a_1\alpha + \cdots + a_{n-1}\alpha^{n-1} + \alpha^n = 0$$

for suitable $a_i \in K$. If [,] denotes the Lie product, then taking the Lie product of the above expression with β yields

$$a_1[\alpha, \beta] + a_2[\alpha^2, \beta] + \cdots + a_{n-1}[\alpha^{n-1}, \beta] + [\alpha^n, \beta] = 0.$$

Now we take the Lie product of this result with $[\alpha, \beta]$ and obtain

$$a_2[[\alpha^2, \beta], [\alpha, \beta]] + \cdots + a_{n-1}[[\alpha^{n-1}, \beta], [\alpha, \beta]] + [[\alpha^n, \beta], [\alpha, \beta]] = 0.$$

Next we take the Lie product with $[[\alpha^2, \beta], [\alpha, \beta]]$ and we continue this process n times until all the coefficients a_i are deleted from the expression. In this way we obtain a specific polynomial identity in two variables satisfied by E. It is easily seen to be nontrivial.

Theorem 28.3. If $K[G]$ is an algebraic algebra of bounded degree, then G is locally finite.

Proof. If $x \in G$, then x is algebraic over K and hence clearly x has finite order. Therefore G is a periodic group. Moreover by Lemma 28.2, $K[G]$ satisfies a polynomial identity and hence by Theorem 5.5, $[G:\Delta(G)] < \infty$. Let H be a finitely generated subgroup of G. Then $[H:H \cap \Delta(G)] < \infty$ and by Lemma 6.1, $H \cap \Delta(G)$ is a finitely generated periodic subgroup of $\Delta(G)$. It follows easily from Lemma 2.2 that $H \cap \Delta(G)$ is finite and the result follows.

If $K[G]$ is algebraic, then as we saw above G is a periodic group and hence every finitely generated subgroup of G is a finitely generated periodic group. The general Burnside problem asks whether such groups must necessarily be finite. The following result settles this in the negative.

Theorem 28.4 (Golod [*13*]). Let K be a countable field of characteristic $p > 0$. Then there exists a finitely generated infinite p-group G and a homomorphic image R of $K[G]$ such that $R/JR \simeq K$ and JR is nil but not nilpotent.

Proof. Let $T = K[\zeta_1, \zeta_2, \ldots, \zeta_d]$ be the polynomial ring over K in the d noncommuting variables $\zeta_1, \zeta_2, \ldots, \zeta_d$ and choose some $d \geq 3$. Then in the notation of Theorem 27.4 there exists an ideal A of T such that M/A is an infinite dimensional nil algebra. Set $R = T/A$. Since

$$(T/A)/(M/A) \simeq T/M \simeq K$$

it follows that $M/A = JR$, $R/JR \simeq K$, and JR is nil and infinite dimensional over K. Hence JR is not nilpotent since otherwise $A \supseteq T_n$ for all sufficiently large n, a contradiction.

Let $\bar{U}$ be the subset of R given by $\bar{U} = \{1 + \alpha \mid \alpha \in JR\}$. Then certainly $\bar{U}$ is closed under multiplication. Let $1 + \alpha \in \bar{U}$. Since JR is nil, α is nilpotent and hence $\alpha^{p^n} = 0$ for some sufficiently large n. Since K has characteristic p, this yields

$$(1 + \alpha)^{p^n} = 1 + \alpha^{p^n} = 1$$

and hence every element of $\bar{U}$ is a unit of finite order. Therefore $\bar{U}$ is a periodic group and in fact a p-group.

Let $\bar{G}$ be the finitely generated subgroup of $\bar{U}$ generated by $\bar{x}_1 = 1 + \bar{\zeta}_1$, $\bar{x}_2 = 1 + \bar{\zeta}_2, \ldots, \bar{x}_d = 1 + \bar{\zeta}_d$ where $\bar{\zeta}_i$ is the image in R of $\zeta_i \in T$. Now let G be an isomorphic copy of $\bar{G}$ with corresponding generators $x_1, x_2, \ldots, x_d$. Then G is a finitely generated p-group and we have an algebra homomorphism $K[G] \rightarrow R$ given by $x_i \rightarrow \bar{x}_i$. Since the image of $K[G]$ is an algebra containing $1, 1 + \bar{\zeta}_1, 1 + \bar{\zeta}_2, \ldots, 1 + \bar{\zeta}_d$, it follows that this map is onto. Since R is infinite dimensional over K, G is infinite and the result follows.

Corollary 28.5. There exists a finitely generated p-group G such that $K[G]$ is prime for any field K.

Proof. By Theorem 28.4 there exists a finitely generated infinite p-group H. Let Z_p be cyclic of order p and set $G = Z_p \wr H$. By Lemma 21.5(ii)(iv), G is a p-group with no nontrivial finite normal subgroups and hence $K[G]$ is prime for all K by Theorem 2.5.

Now G has a normal subgroup W which is a direct product of copies of Z_p, indexed by the elements of H and $G = WH$. Moreover it is easy to see that G is generated by H and any one of these copies of Z_p. Therefore G is finitely generated.

We close this section with an amusing example. The following lemma is due to Amitsur.

Lemma 28.6. Let E be an algebra over K and let F be a field extension of K with infinite transcendence degree. Then $J(F \otimes_K E)$ is a nil ideal.

Proof. Let $\alpha \in J(F \otimes E)$. Then clearly $\alpha \in L \otimes E$ for some field L with $F \supseteq L \supseteq K$ and t.d. $(L/K) < \infty$. Now t.d. $(F/K) = \infty$ so there exists a second field L' with $F \supseteq L' > L$ and with L'/L purely transcendental. If we set $A = L \otimes E$, then by Lemma 16.11 we have $\alpha \in J(L' \otimes_L A) \cap A$ and hence α is nilpotent by Theorem 17.10.

Lemma 28.7. Let G be a group and let K be a field of infinite transcendence degree over its prime subfield K_0. If $K[G] = K + JK[G]$, then $K[G]$ is algebraic.

Proof. Since $K[G] = K \otimes_{K_0} K_0[G]$, Lemma 28.6 implies that $JK[G]$ is nil. Now if $\alpha \in K[G]$, then by assumption there exists $a \in K$ with $\alpha - a \in JK[G]$. Thus $(\alpha - a)^n = 0$ for some n and α is algebraic over K.

Finally let G be the finitely generated infinite p-group of Theorem 28.4 and let K be a field of characteristic p with t.d. $\big(K/GF(p)\big) = \infty$. Note that K could still be countable under this assumption. If $K[G] = K + JK[G]$, then by the previous lemma, $K[G]$ is algebraic but G is not locally finite. On the other hand if $K[G] \neq K + JK[G]$, then we have found a p-group in characteristic p such that $JK[G]$ is not equal to

$$\{\sum a_x \cdot x \in K[G] \mid \sum a_x = 0\}$$

which is surprising in view of Lemma 21.5(i) and Theorem 28.4. Thus one of these two cases yields an interesting example but we do not know which one.

V

RESEARCH PROBLEMS

The algebraic study of group rings is a fairly new subject and there are numerous unsolved problems. Some of these are listed below with a discussion and sketchy proofs of some pertinent lemmas. To begin, let us consider polynomial identities.

Problem 1. Find necessary and sufficient conditions for $K[G]$ to satisfy a polynomial identity.

This has of course been solved in characteristic 0. In characteristic p, we have

Lemma. Suppose G has a normal subgroup A of finite index such that A' is a finite p-group. If K has characteristic p, then $K[G]$ satisfies a polynomial identity of degree $n = 2[G:A] \cdot |A'|$.

Proof. Since A' is a finite normal p-subgroup of G, it follows that I, the kernel of the natural map $K[G] \to K[G/A']$, is equal to $JK[A'] \cdot K[G]$ and hence $I^a = 0$ where $a = |A'|$. Now G/A' has an abelian subgroup A/A' of index $b = [G:A]$ and hence $K[G/A']$ satisfies a polynomial identity f of degree $2b$. It follows easily that $K[G]$ satisfies f^a, a polynomial identity of degree $2ab$.

A group A is said to be p-abelian if A' is a finite p-group. In view of the above lemma we can propose a more specific version of (1).

Problem 2. Let K be a field of characteristic p and let $K[G]$ satisfy a polynomial identity of degree n. Does G necessarily have a normal p-abelian subgroup A of finite index? If this is the case, are $[G:A]$ and $|A'|$ necessarily bounded as functions of n?

An affirmative answer to (2) would imply easily that $\mathbf{S}(G)$ is solvable of bounded deiived length. Therefore we might pose the simpler

Problem 3. Let K be a field of characteristic p and let $K[G]$ satisfy a polynomial identity of degree n. Is $\mathbf{S}(G)$ solvable of bounded derived length?

It is amusing that the study of (3) is equivalent to the special case of finite p-groups.

Lemma. An affirmative answer to (3) in the case of finite p-groups implies an affirmative answer in general.

Proof. We may assume throughout that $K[G]$ satisfies a polynomial identity of degree n and we may clearly suppose that K is an algebraically closed field of characteristic p. If H is a solvable group, let $d(H)$ denote its derived length. Then we are given a function $\delta_1(n)$ such that if G is a finite p-group, then $d(G) \leq \delta_1(n)$.

Suppose first that G is a finite group and that $K[G]$ has r.b. 1. Then G' is a p-group so $d(G') \leq \delta_1(n)$ and hence $d(G) \leq \delta_1(n) + 1$. Now let G be a finite solvable group with derived series $G = G^0 \supseteq G^1 \supseteq G^2 \supseteq \cdots$ and set $H_i = G^{i(\delta_1+2)}$. Say $G = H_0 > H_1 > \cdots > H_s > H_{s+1} = \langle 1 \rangle$. Then for $1 \leq i \leq s$, $d(H_{i-1}/H_i) = \delta_1 + 2$ and thus $K[H_{i-1}/H_i]$ does not have r.b. 1. Now $K[H_0]$ has r.b. $[n/2]$ so $K[H_1]$ has r.b. $[n/4]$, $K[H_2]$ has r.b. $[n/8]$, and continuing in this manner we see that $K[H_s]$ has r.b. $[n/2^{s+1}]$. Thus $s + 1 \leq n$ and $d(G) \leq \delta_2(n)$ where $\delta_2(n) = \big(\delta_1(n) + 2\big) \cdot n$.

Suppose now that G is finitely generated and solvable. Then G has a normal abelian subgroup A with $|G/A|$ finite. Thus $d(G/A) \leq \delta_2(n)$ and $d(G) \leq \delta_2(n) + 1$. We now consider the general case. It clearly suffices to assume that $G = \mathbf{S}(G)$. Then G is locally solvable and each such finitely generated solvable subgroup has bounded derived length. This clearly implies that G is solvable with $d(G) \leq \delta_2(n) + 1$ and the result follows.

From an esthetic point of view we might ask

Problem 4. Let $K[G]$ satisfy a polynomial identity of degree n.

(i) Is $[G:\Delta(G)] \leq n/2$ in all cases?
(ii) What are the best possible bounds for $[G:\mathrm{S}(G)]$ and for $[G:A]$ in case $K[G]$ is semiprime?

Problem 5. Is there a purely combinatorial proof, that is one that does not use representation degrees, of Theorem 13.4?

We can in fact give such a proof for the result that $[G:\mathrm{S}(G)] \leq n!\mathfrak{A}(n)$. It requires the following

Lemma. Let N and C be finite groups and let K be algebraically closed. Suppose that $K[N \times C]$ satisfies a polynomial identity of degree n and that $K[C]$ does not have r.b. 1. Then $K[N]$ satisfies a polynomial identity of degree $n - 1$.

Proof. By assumption $K[C]/JK[C]$ contains as a subring K_2, the ring of 2×2 matrices over K. It then follows easily that $(K[N])_2$, the ring of 2×2 matrices over $K[N]$, is a subring of a quotient of $K[N \times C]$. Hence $(K[N])_2$ satisfies a polynomial identity f of degree n.

We may assume that

$$f(\zeta_1, \zeta_2, \cdots, \zeta_n) = \zeta_1\zeta_2 \ldots \zeta_n + \sum_{\sigma \neq 1} a_\sigma \zeta_{\sigma(1)}\zeta_{\sigma(2)} \cdots \zeta_{\sigma(n)}$$

and define $g(\zeta_2, \ldots, \zeta_n)$ by

$$f(\zeta_1, \zeta_2, \ldots, \zeta_n) = \zeta_1 g(\zeta_2, \ldots, \zeta_n) + \textit{terms not starting with } \zeta_1.$$

Let $\alpha_2, \alpha_3, \ldots, \alpha_n \in K[N]$ and put $\zeta_1 = e_{21}$ and $\zeta_i = e_{11}\alpha_i$ for $i > 1$ in f. Here e_{11} and e_{21} are the usual matrix units in K_2. We then have easily

$$0 = f(\zeta_1, \zeta_2, \ldots, \zeta_n) = e_{21} g(\alpha_2, \ldots, \alpha_n)$$

and hence $K[N]$ satisfies the polynomial identity g of degree $n - 1$.

Now we merely observe that in the proof of Theorem 12.4 we use the r.b. n argument to show that $K[N]$ has some smaller parameter than $K[N \times C]$.

If K is algebraically closed and has characteristic 0 and if G is a finite p-group, then all irreducible representations of $K[G]$ have degrees which are powers of p. If the biggest such degree is p^e then we let $e = e(G)$ be the representation exponent of G. In the notation of Theorem 8.3, it was shown that $|G_{p^{e-1}}| \leq (2p^{e-1})!$.

Problem 6. Let K, G, and $G_{p^{e-1}}$ be as above. Is $|G_{p^{e-1}}|$ necessarily bounded by a function of e independent of p?

Now all nonidentity subgroups of G have order at least p. Thus a consequence of (6) would be

Problem 7. Let K, G, and $G_{p^{e-1}}$ be as above. Does there exist a function $\eta(e)$ of e such that $p > \eta(e)$ implies that $G_{p^{e-1}} = \langle 1 \rangle$? If so, is $\eta(e) = e$?

An affirmative answer to (7) would have interesting applications to the study of p-groups G as a function of their representation exponent. The result is known to be true in a number of special cases, including the case of groups of nilpotence class 2, and here $\eta(e) = e$ is the best possible function (see [*21*]).

There are apparently analytic proofs of some special cases of the following. We pose the general problem.

Problem 8. Let $K[G]$ be semisimple and suppose that all its irreducible representations have finite degree (defined as in Section 8). Must $K[G]$ necessarily have r.b. n for some n?

Let K be a field and let G be a group. Then a twisted group ring $K^t[G]$ of G over K is an associative K-algebra with basis $\{\bar{x} \mid x \in G\}$ and with multiplication defined distributively by $\bar{x}\bar{y} = \gamma(x, y)\overline{xy}$ for $x, y \in G$ and for some $\gamma(x, y) \in K - \{0\}$. The associativity condition is equivalent to $\bar{x}(\bar{y}\bar{z}) = (\bar{x}\bar{y})\bar{z}$ for all $x, y, z \in G$ and this is equivalent to

$$\gamma(x, yz)\gamma(y, z) = \gamma(x, y)\gamma(xy, z).$$

If $\gamma(x, y) = 1$ for all x, y, then clearly $K^t[G] = K[G]$. These twisted group rings occur in a rather natural way.

Lemma. Let $K[G]$ be a group ring, let Z be a central subgroup of G, and let I be an ideal in $K[Z]$ with $K[Z]/I \simeq K$. Then $I \cdot K[G]$ is an ideal in $K[G]$ and $K[G]/I \cdot K[G] \simeq K^t[G/Z]$ where the latter is some twisted group ring of G/Z.

Proof. Since Z is central in G, we have $I \cdot K[G] = K[G] \cdot I$ so this is an ideal in $K[G]$. For each $x \in G/Z$, let $\bar{x} \in G$ be a fixed inverse image. It then follows easily that $K[G]/I \cdot K[G]$ is an associative K-algebra with basis $\{\bar{x} \mid x \in G/Z\}$. Moreover if $x, y \in G/Z$, then $\bar{x}\bar{y} = z\overline{xy}$ for some

$z \in Z$ so in $K[G]/I \cdot K[G]$ we have $\bar{x}\bar{y} = \gamma(x, y)\overline{xy}$ where $\gamma(x, y) \in K - \{0\}$ is the image of z in $K[Z]/I = K$.

In view of the above, it appears useful to the study of ordinary group rings to also study twisted group rings. We offer the following blanket

Problem 9. Study twisted group rings and find the appropriate analogs for all the results proved here.

Actually most of the results of Chapter III are also known for twisted group rings (see [*40*]) and many results on polynomial identities can be obtained simply (see [*45*]) from the ordinary case. If R is a ring, we let R^0 denote the opposite ring of R. Thus $R = R^0$ as sets and addition is the same. Moreover for $\alpha, \beta \in R^0$ we have $\alpha \circ \beta = \beta\alpha$.

Lemma. Let $K^t[G]$ be a twisted group ring. Then $K[G]$ is K-isomorphic to a subalgebra of $K^t[G] \otimes_K K^t[G]^0$.

Proof. We first observe that $K^t[G]$ has a 1. Put $y = 1$ in the associativity condition for γ. Then we have $\gamma(1, z) = \gamma(x, 1)$ and thus for all $x, z \in G$, $\gamma(1, z) = \gamma(x, 1) = \gamma(1, 1)$. Therefore $\gamma(1, 1)^{-1}\bar{1}$ is the identity of $K^t[G]$. Now for $x \in G$, $\bar{x}\overline{x^{-1}}$ and $\overline{x^{-1}}\bar{x}$ are scalar multiples of 1. Thus $\bar{x}$ has an inverse which is a scalar multiple of $\overline{x^{-1}}$.

Let $x, y \in G$. In $K^t[G]$ we have $\bar{x}\bar{y} = \gamma(x, y)\overline{xy}$ so taking inverses yields $\bar{y}^{-1}\bar{x}^{-1} = \gamma(x, y)^{-1}\overline{xy}^{-1}$. Thus in $K^t[G]^0$ we have $\bar{x}^{-1} \circ \bar{y}^{-1} = \gamma(x, y)^{-1}\overline{xy}^{-1}$. Finally in $K^t[G] \otimes K^t[G]^0$ we have

$$\begin{aligned}(\bar{x} \otimes \bar{x}^{-1})(\bar{y} \otimes \bar{y}^{-1}) &= (\bar{x}\bar{y}) \otimes (\bar{x}^{-1} \circ \bar{y}^{-1}) \\ &= \gamma(x, y)\gamma(x, y)^{-1}\overline{xy} \otimes \overline{xy}^{-1} \\ &= \overline{xy} \otimes \overline{xy}^{-1}.\end{aligned}$$

This shows that the K-linear map $K[G] \to K^t[G] \otimes K^t[G]^0$ given by $x \to \bar{x} \otimes \bar{x}^{-1}$ is an injective isomorphism.

Now suppose that $K^t[G]$ satisfies a polynomial identity. Then so does $K^t[G]^0$ and also $K^t[G] \otimes K^t[G]^0$ if $K^t[G]$ is semisimple. Therefore by the above lemma $K[G]$ satisfies a polynomial identity. A strengthening of Corollary 8.2 is therefore needed to handle the general case.

Problem 10. Let E_1 and E_2 be K-algebras satisfying polynomial identities. Does $E_1 \otimes E_2$ necessarily satisfy a polynomial identity?

If E_1 and E_2 can be embedded in full matrix rings over commutative algebras, then the result is trivial. However we know by Theorem 8.3 that this is not always the case. It may nevertheless be true for group rings.

Problem 11. Let $K[G]$ satisfy a polynomial identity of degree n. Can $K[G]$ be embedded in a full matrix ring over some commutative K-algebra? If so, what is the minimal possible degree of the matrix ring?

Presumably an affirmative answer to (2) will yield an affirmative answer here. We therefore restate the above to read

Problem 12. Let G have a p-abelian subgroup A of finite index and let K be a field of characteristic p. Can $K[G]$ be embedded in a full matrix ring over a commutative K-algebra?

There are a number of simple techniques which reduce the above to a more manageable problem. We proceed with a series of lemmas. All rings are K-algebras and all tensor products are over K. The following is just the usual induced representation. The third lemma is a generalization of a result of Burnside on faithful representations of finite groups.

Lemma. Let $[G:H] = m < \infty$ and suppose that $\rho: K[H] \to R$ is injective. Then there exists an injective homomorphism $\rho^*: K[G] \to R_m$.

Proof. As in the proof of Theorem 5.1 we have $K[G] \subseteq (K[H])_m$. Thus there is clearly a natural injection $\rho^*: K[G] \to R_m$.

Lemma. Let H be a normal subgroup of G. Suppose $\sigma: K[G/H] \to S$ is injective and suppose that $\tau: K[G] \to T$ is injective when restricted to $K[H]$. Then viewing σ as a homomorphism from $K[G]$ to S in the natural way, we have $\sigma \otimes \tau: K[G] \to S \otimes T$ is injective.

Proof. Observe that $\sigma \otimes \tau$ is defined linearly by $(\sigma \otimes \tau)(x) = \sigma(x) \otimes \tau(x)$ for $x \in G$. Let $\{z_i\}$ be a set of coset representatives for H in G. If $x \in H$, then

$$(\sigma \otimes \tau)(xz_i) = \sigma(xz_i) \otimes \tau(xz_i) = \sigma(z_i) \otimes \tau(x)\tau(z_i).$$

Let $\alpha \in K[G]$ and write $\alpha = \sum \alpha_i z_i$ with $\alpha_i \in K[H]$. Then the above yields

$$(\sigma \otimes \tau)(\alpha) = \sum \sigma(z_i) \otimes \tau(\alpha_i)\tau(z_i).$$

Suppose now that $(\sigma \otimes \tau)(\alpha) = 0$. Since σ is injective on $K[G/H]$, it follows that the elements $\sigma(z_i)$ are K-linearly independent. Thus we must have $\tau(\alpha_i)\tau(z_i) = 0$ for all i and hence $\tau(\alpha_i) = 0$ since $\tau(z_i)$ is invertible. Now τ is assumed to be injective when restricted to $K[H]$ so $\tau(\alpha_i) = 0$ implies $\alpha_i = 0$ and hence $\alpha = 0$. Thus $\sigma \otimes \tau$ is injective.

Lemma. Let $\sigma: K[G] \to S$ and $\tau: K[G] \to T$ be two homomorphisms and let s and t be fixed positive integers. Suppose that the image under σ of every subset of G of size $\leq s$ is K-linearly independent and that the image under τ of every subset of G of size $\leq t$ is K-linearly independent. Then the image under $\sigma \otimes \tau: K[G] \to S \otimes T$ of every subset of G of size $\leq s + t - 1$ is K-linearly independent.

Proof. Let U be a finite subset of G such that $(\sigma \otimes \tau)(U)$ is dependent and write

$$0 = \sum_{x \in U} c_x(\sigma \otimes \tau)(x)$$

with $c_x \in K$ and some c_x not zero, say $c_{x_1} \neq 0$. Write $U = \{x_1, x_2, \ldots, x_n, y_1, y_2, \ldots, y_m\}$ so that $\sigma(x_1), \sigma(x_2), \ldots, \sigma(x_n)$ are linearly independent and that each $\sigma(y_i)$ is dependent on these. Note that as yet we do not know that the y's exist. Say $\sigma(y_i) = \sum_j d_{ij}\sigma(x_j)$. Then the above dependence relation yields easily

$$0 = \sum_i \sigma(x_i) \otimes \left(c_{x_i}\tau(x_i) + \sum_{j=1}^{m} c_{y_j} d_{ji}\tau(y_j)\right).$$

Since the $\sigma(x_i)$ are linearly independent, this yields for $i = 1$

$$0 = c_{x_1}\tau(x_1) + \sum_{j=1}^{m} c_{y_j} d_{j1}\tau(y_j)$$

so since $c_{x_1} \neq 0$ we must have $1 + m > t$ by assumption. In particular $m > 1$. Then $\sigma(x_1), \sigma(x_2), \ldots, \sigma(x_n), \sigma(y_1)$ are dependent so we have $1 + n > s$ by assumption. Thus $m \geq t, n \geq s$, and $|U| = m + n \geq s + t$.

These can now be combined to yield our final reduction. We first state a problem which appears off hand to be simpler than (12).

Problem 13. Let G be a group whose commutator subgroup G' is a finite cyclic p-group central in G and let K be a field of characteristic p. Must there necessarily exist a homomorphism $\sigma: K[G] \to C_n$ where C is a commutative K-algebra such that σ is faithful on G'.

Lemma. An affirmative answer to (13) implies an affirmative answer to (12).

Proof. Suppose (13) is known to be true and let G be a group with a p-abelian subgroup A of finite index in G. We apply the previous three lemmas in turn to obtain our reduction.

Suppose H is a subgroup of G of finite index m and suppose $\rho: K[H] \to C_n$ is injective. Then by the first lemma after (12) there exists an injection $\rho^*: K[G] \to (C_n)_m = C_{nm}$. Thus to prove (12) for G it suffices to prove it for any subgroup of finite index. In particular we can assume that G is p-abelian. Let $L = \mathbf{C}_G(G')$. Since G' is finite, we have $[G:L] < \infty$. Moreover $L' \subseteq G'$ so L centralizes L' and L' is a finite central p-subgroup of L. Thus we may also assume that $G = L$ or equivalently that G' is a finite central p-subgroup of G.

The second lemma has two applications to the problem. First suppose that $\bar{G} = H_1 \times H_2$ and suppose that $\rho_1: K[H_1] \to (C^1)_{n_1}$, $\rho_2: K[H_2] \to (C^2)_{n_2}$ are injections with C^1 and C^2 commutative. Then we can view ρ_1 and ρ_2 as homomorphisms from $K[\bar{G}]$ and by the second lemma $\rho_1 \otimes \rho_2: K[\bar{G}] \to (C^1 \otimes C^2)_{n_1 n_2}$ is an injection. Thus if $K[H_1]$ and $K[H_2]$ are embeddable in matrix rings over commutative rings, then so is $K[H_1 \times H_2]$. By induction, this holds for any finite number of factors.

Now G' is a finite central p-subgroup of G so there exists a finite collection of subgroups $N_i \subseteq G'$ with G'/N_i cyclic and with $\bigcap N_i = \langle 1 \rangle$. Clearly the natural map $G \to \prod_i G/N_i$ is injective. Thus by the above it suffices to show that each $K[G/N_i]$ is contained in such a C_n or equivalently we can assume that G' is a finite central cyclic p-subgroup of G.

Now suppose $\tau: K[G] \to C_n$ is injective when restricted to $K[G']$ and suppose that C is commutative. Then $\sigma: K[G/G'] \to K[G/G']$, the identity map, is certainly an injection. So again by the second lemma $\sigma \otimes \tau: K[G] \to (K[G/G'] \otimes C)_n$ is injective. Since $K[G/G'] \otimes C$ is commutative we see that it suffices to find a homomorphism $\rho: K[G] \to C_n$ where C is commutative which is an injection when restricted to $K[G']$. We now find such a homomorphism.

Let $\sigma: K[G] \to C_n$ be the homomorphism given by (13) and for each integer s define $\sigma_s: K[G] \to C_n \otimes C_n \otimes \cdots \otimes C_n$ (s times) by $\sigma_s = \sigma \otimes \sigma \otimes \cdots \otimes \sigma$. We claim that the image under σ_s of every subset of G' of size $\leq s + 1$ is K-linearly independent. Consider $s = 1$, so $\sigma_1 = \sigma$. Let $x, y \in G'$ and suppose that $\sigma(x)$ and $\sigma(y)$ are dependent. Since $\sigma(1) = 1$ it follows that $\sigma(xy^{-1}) \in K$. Now xy^{-1} has order a power of p and K has characteristic p. Thus $\sigma(xy^{-1}) = 1$ so $\sigma(x) = \sigma(y)$. Since σ is faithful on G' this yields $x = y$ and the $s = 1$ case follows. Finally $\sigma_{s+1} = \sigma_s \otimes \sigma_1$ so the general case follows from the third lemma after (12) by induction, by restricting these homomorphisms down to $K[G']$.

Thus if $s = |G'| - 1$ then all elements of $\sigma_s(G')$ are linearly independent so σ_s is injective when restricted to $K[G']$. Since $\sigma_s: K[G] \to D_m$ where $D = C \otimes C \otimes \cdots \otimes C$ (s times) is commutative and where $m = n^s$, the result follows.

This of course is all academic unless we can actually construct some examples for (13). One can in fact handle the case in which G' is cyclic and central of order p. Strangely enough the center of the group seems to cause trouble so we will just consider an easier special case. Roughly speaking we assume that $G' = \langle z \rangle$ is cyclic and central of order p and that G/G' is an elementary abelian p-group freely generated by the images of $x_1, x_2, x_3, \ldots \in G$ (not necessarily countable) and moreover $x_i^p = 1$. The following result is due to D. S. Birkes (unpublished).

Lemma. Let K be a field of characteristic p and let

$$G = \langle z, x_1, x_2, x_3, \ldots \mid z^p = 1, \quad (z, x_i) = 1, \quad x_i^p = 1, \quad (x_i, x_j) = z^{f(i,j)} \rangle.$$

Then there exists a commutative K-algebra C and a homomorphism $\sigma: K[G] \to C_2$ such that σ is faithful on G'.

Proof. Observe that $f(i, j)$ is just a function into the integers modulo p. Of course $f(i, i) = 0$ and also $f(i, j) = -f(j, i)$ since in any group $(y, x) = (x, y)^{-1}$. Thus we can assume that the subscripts are linearly ordered somehow and we only need the relations $(x_i, x_j) = z^{f(i,j)}$ for $i > j$.

Now define a K-algebra S as follows. Note that here S is not assumed to be associative, nor does it have a 1. S has a K-basis $\{e, a_1, a_2, a_3, \ldots, b_1, b_2, b_3, \ldots\}$ where the subscripts are the same as those for the x_i and multiplication is defined by

$$0 = e^2 = ea_i = a_ie = eb_i = b_ie$$

$$0 = a_ia_j = b_ib_j$$

$$a_ib_j = b_ja_i = \begin{matrix} f(i,j)e & \textit{for} & i > j \\ 0 & \textit{for} & i \leq j. \end{matrix}$$

Observe that S is commutative and also associative since any product of three basis elements of S is zero. Let C equal S with a 1 adjoined so that C is the vector space direct sum $C = S + K$ with the obvious multiplication.

Consider the matrix ring C_2 and define matrices

$$X_i = \begin{pmatrix} 1 & a_i \\ b_i & 1 \end{pmatrix} = \begin{pmatrix} 1 & 0 \\ 0 & 1 \end{pmatrix} + \begin{pmatrix} 0 & a_i \\ b_i & 0 \end{pmatrix}$$

$$Z = \begin{pmatrix} 1+e & 0 \\ 0 & 1-e \end{pmatrix} = \begin{pmatrix} 1 & 0 \\ 0 & 1 \end{pmatrix} + \begin{pmatrix} e & 0 \\ 0 & -e \end{pmatrix}.$$

Since $a_i b_i = b_i a_i = 0$ we have

$$\begin{pmatrix} 0 & a_i \\ b_i & 0 \end{pmatrix}^2 = 0$$

so clearly $X_i^p = 1$ since K has characteristic p. Also $e^2 = 0$ so $Z^p = 1$. Moreover Z commutes with each X_i since e annihilates a_i and b_i. Finally we compute the commutators (X_i, X_j) for $i > j$. Now $\det X_i = 1 - a_i b_i = 1$ so

$$X_i^{-1} = \begin{pmatrix} 1 & -a_i \\ -b_i & 1 \end{pmatrix}.$$

This yields easily

$$(X_i, X_j) = X_i^{-1} X_j^{-1} X_i X_j = \begin{pmatrix} 1 + a_i b_j - a_j b_i & 0 \\ 0 & 1 + a_j b_i - a_i b_j \end{pmatrix}$$

$$= \begin{pmatrix} 1 & 0 \\ 0 & 1 \end{pmatrix} + f(i,j) \begin{pmatrix} e & 0 \\ 0 & -e \end{pmatrix}$$

$$= Z^{f(i,j)}$$

since $e^2 = 0$.

Thus the map $\sigma: K[G] \to C_2$ defined by $x_i \to X_i$, $z \to Z$ yields a representation for $K[G]$ and since $Z \neq 1$ this is faithful on G'.

Without a doubt the most difficult group ring problems concern the question of semisimplicity. We start with the hardest of these.

Problem 14. Find necessary and sufficient conditions for $K[G]$ to be semisimple.

Problem 15. Describe $JK[G]$ in general. Is it always a nil ideal?

The first part of (15) is not even known for finite groups and is therefore presumably a hopeless task. The second part is of course known to be the case if K has infinite transcendence degree over its prime subfield. Now let $\alpha \in JK[G]$. Then there exists a finitely generated subgroup H such that $\alpha \in JK[G] \cap K[H] \subseteq JK[H]$. Since $K[H]$ is clearly a finitely generated algebra, the latter part of (15) would follow from an affirmative answer to the next question which is a noncommutative analog of the Hilbert Nullstellensatz.

Problem 16. Let E be a finitely generated K-algebra. Is JE necessarily a nil ideal?

There are a number of interesting special cases of (14) which should be considered. Perhaps the two most important are

Problem 17. Suppose that G has no elements of order p in case K has characteristic p. Show that $K[G]$ is semisimple.

Problem 18. Study the semisimplicity question for p-groups in characteristic p. Can $K[G]$ have nontrivial irreducible representations or must we always have $K[G]/JK[G] \simeq K$?

We of course know (17) to be true if K is not algebraic over its prime subfield. Thus since the Jacobson radical behaves well under separably algebraic field extensions, the only cases which need be considered are $K = Q$, the rationals, and $K = GF(p)$. For (18), Golod–Shafarevitch techniques may yield interesting examples. Some additional special cases of interest are

Problem 19. Let S_∞ denote the infinite symmetric group consisting of all permutations moving only finitely many points. Is $K[S_\infty]$ semisimple for all fields?

Problem 20. Let G be a solvable group. Find necessary and sufficient conditions for $K[G]$ to be semisimple.

If G is metabelian the answer to (20) is known (see [*38*], [*43*]) to be: $JK[G] \neq 0$ if and only if G has a finite subgroup H whose order is divisible by the characteristic of K such that $\mathbf{N}_G(H)$ is normal in G and $G/\mathbf{N}_G(H)$ is locally finite. Apparently nothing is known for groups of derived length greater than 2. We might also wish to consider the family of nilpotent groups but here the result is trivial.

Lemma. Let $\langle 1 \rangle \subseteq Z_1 \subseteq Z_2 \subset \cdots$ denote the upper central series of G and let p be a fixed prime. If Z_1 has no elements of order p, then neither does Z_{i+1}/Z_i for all $i \geq 1$.

Proof. By induction on i, it clearly suffices to show that Z_2/Z_1 has no elements of order p.

Let $x \in G$. We first observe that the commutator map $a \to (a, x)$ is a homomorphism from Z_2 to Z_1. If $a \in Z_2$ then certainly $(a, x) \in Z_1$. Let $a, b \in Z_2$. Then

$$\begin{aligned}(ab, x) &= (ab)^{-1}(ab)^x = b^{-1}(a^{-1}a^x)b^x \\ &= (a^{-1}a^x)(b^{-1}b^x) = (a, x)(b, x).\end{aligned}$$

Now let $a \in Z_2$ and suppose that $a^p \in Z_1$. If $x \in G$ then by the above we have $1 = (a^p, x) = (a, x)^p$ and hence $1 = (a, x)$ since Z_1 has no elements of order p. Thus $a \in Z_1$ and Z_2/Z_1 has no elements of order p.

Now let G be nilpotent. If either K has characteristic 0 or K has characteristic p and G has no elements of order p, then since G is solvable we have $JK[G] = 0$. The remaining case is that K has characteristic p and that G has an element of order p. Since G is nilpotent some quotient in the upper central series of G will have an element of order p and hence by the above lemma $\mathbf{Z}(G)$ has an element x of order p. It follows that $(1 - x)K[G]$ is a nontrivial nilpotent ideal so $JK[G] \neq 0$.

Let K be a field. In [*58*] Wallace defines a group G to be a JK-group if for all groups H and normal subgroups W, $H/W \simeq G$ implies that $JK[H] \subseteq (JK[W]) \cdot K[H]$. If G is a JK-group, then applying the above with $H = G$, $W = \langle 1 \rangle$, we obtain $JK[G] = 0$. Thus we ask

Problem 21. Is G a JK-group if and only if $JK[G] = 0$?

This is of course true if G is finite or abelian by the results of Sections 16 and 17. Moreover the analogous result for the nilpotent radical is true by Corollary 20.4. Again some special cases might be worth considering.

Problem 22. Let G be an infinite dihedral group, that is, $G = \langle x, y \mid y^2 = 1, y^{-1}xy = x^{-1} \rangle$. Is G a JK-group for any field K of characteristic 2?

Suppose $H/W = G$ is an infinite dihedral group and let $H > H_0 > W$ with $H_0/W = \langle x \rangle$. Then we know that $JK[H_0] \subseteq (JK[W]) \cdot K[H_0]$ so the difficulty arises in going from H_0 to H and the following would be helpful.

Problem 23. Let G be a group with a subgroup G_0 of index 2. If K is a field of characteristic 2, study closely the relationship between $JK[G]$ and $JK[G_0]$.

Suppose now that W is a finite central subgroup of H whose order is prime to the characteristic of K and let $H/W = G$. Suppose further that K is algebraically closed. Since $K[W]$ is semisimple, there exist centrally primitive idempotents $e_i \in K[W]$ such that $K[W] = e_1K[W] + \cdots + e_nK[W]$ is a ring direct sum and $e_iK[W] = e_iK$. Since W is central in H we have

$$K[H] = e_1K[H] + e_2K[H] + \cdots + e_nK[H]$$

a ring direct sum and hence

$$JK[H] = J(e_1K[H]) + J(e_2K[H]) + \cdots + J(e_nK[H]).$$

Now by the lemma preceding (9), $e_iK[H]$ is some twisted group ring $K^{t_i}[G]$ of $H/W = G$. Thus we see that even if $JK[G] = 0$, G will not be a JK-group unless $JK^{t_i}[G] = 0$ for all i.

Problem 24. Let K be a perfect field and let G be a group. Is the statement $JK^t[G] = 0$ independent of the twisting, that is, of the cofactor system $\{\gamma(x, y)\}$?

This is known to be true for the nilpotent radical (see [*40*]). We remark that the assumption above that K is perfect is necessary. For suppose that K has characteristic p and that $a \in K$ does not have a pth root in K. Then the polynomial $\zeta^p - a$ is irreducible over K so $K[\zeta]/(\zeta^p - a)$ is a field. On the other hand, this quotient is easily seen to be a twisted group ring $K^t[Z_p]$ of Z_p, the cyclic group of order p. Thus $JK[Z_p] \neq 0$ but $JK^t[Z_p] = 0$.

Let E be an algebra over a field K. We say that E is eventually semisimple if E is nilpotent free (see Section 18) and if $J(F \otimes_K E) = 0$ for some field extension F of K. According to Theorem 18.2 the statement $J(F \otimes E) = 0$ is merely a function of the transcendence degree of F/K and we define the degree of E to be equal to the minimum such transcendence degree.

Problem 25. Let E be an eventually semisimple algebra over K. What are the possible values for the degree of E as defined above? Can degrees other than 0, 1, and ∞ occur?

There are a number of unsettled problems of a special nature. We lump three of these together. A ring is primitive if it has a faithful irreducible representation. A ring R is semiperfect (according to Bass) if R/JR is Artinian and if idempotents in the quotient can be lifted to R. It is an elementary exercise to verify that any perfect ıing is also semiperfect.

Problem 26. Find necessary and sufficient conditions for the group ring $K[G]$ to be

(i) primitive
(ii) semiperfect
(iii) algebraic.

Problem 27. Is it true that $K[G]$ is Noetherian if and only if G has a series of subgroups $G = G_0 \supseteq G_1 \supseteq \cdots \supseteq G_n = \langle 1 \rangle$ such that G_{i+1} is normal in G_i and G_i/G_{i+1} is either a finite group or infinite cyclic.

The fact that $K[G]$ is Noetherian if G has the above series is a consequence of induction and the following

Lemma. Let H be a normal subgroup of G and suppose that $K[H]$ is right Noetherian. If G/H is either a finite group or infinite cyclic, then $K[G]$ is right Noetherian.

Proof. Suppose first that G/H is finite. Then $K[G]$ is a finitely generated right $K[H]$-module and thus it is a Noetherian module. Since every right ideal of $K[G]$ is a $K[H]$-submodule, this case follows. Now let G/H be infinite cyclic. Then there exists an element $x \in G$ of infinite order such that $G = H\langle x \rangle$. Thus $K[G]$ looks like a polynomial ring in x, with negative exponents also, over $K[H]$ and the result follows easily by aping the proof of the Hilbert basis theorem.

Then there is the difficult zero divisor problem.

Problem 28. Let G be a torsion free group and let K be any field. Is it true that $K[G]$ has no proper divisors of zero? Are all units in $K[G]$ necessarily trivial?

A number of special cases should be considered. If G is nilpotent, the result is trivial.

Lemma. Let G be a torsion free nilpotent group. Then $K[G]$ has no proper divisors of zero and only trivial units.

Proof. We apply the lemma following (20). Since Z_1 is torsion free, it follows that Z_{i+1}/Z_i is torsion free for all i. Then since G is nilpotent, it has a normal series with torsion free abelian quotients and the result follows.

It is in fact not hard to show that G is an ordered group. First order each factor Z_{i+1}/Z_i. Then for $x \in G$, $x \neq 1$, we say $x > 1$ if for the appropriate i with $x \in Z_{i+1} - Z_i$ we have $xZ_i > 1$ in Z_{i+1}/Z_i.

Problem 29. Study (28) in the special case of solvable groups G.

If $K[G]$ is embeddable in a division ring, then certainly it has no proper divisors of zero.

Problem 30. Find necessary and sufficient conditions for $K[G]$ to be embeddable in a division ring.

Lemma. Let G be torsion free and suppose that $K[G]$ satisfies a polynomial identity. If $K[G]$ has a proper divisor of zero, then it has a nilpotent element. If $K[G]$ has no proper divisors of zero, then it is embeddable in a division ring.

Proof. Since G is torsion free, $K[G]$ is prime and we follow the proof of Theorem 6.5. Let Z denote the center of $K[G]$. Then no nonzero element of Z is a zero divisor in $K[G]$ and $Z^{-1}K[G] \simeq D_m$, the ring of $m \times m$ matrices over some division ring D. If $K[G]$ has proper divisors of zero, then $m > 1$. Hence if $\eta^{-1}\alpha = e_{12}$, the matrix unit, then $\alpha \in K[G]$, $\alpha \neq 0$, and $\alpha^2 = 0$. If $K[G]$ has no proper divisors of zero, then since Z is central, neither does $Z^{-1}K[G]$. Hence $m = 1$ and $K[G] \subseteq D$.

Thus another special case of interest is

Problem 31. Study (28) and (30) in the special case of group rings satisfying a polynomial identity.

Now for a few odds and ends.

Problem 32. Let K be a field of characteristic zero and let $e \in K[G]$ be an idempotent. Is the trace of e necessarily rational? If so, find an algebraic proof of this fact.

Problem 33. Is there a characteristic p analog of (32)? Namely, if K has characteristic p and if $e \in K[G]$ is an idempotent, do we always have $\operatorname{tr} e \in GF(p)$?

Problem 34. Let $\alpha, \beta \in K[G]$ with $\alpha\beta = 1$. Must $\beta\alpha$ necessarily also equal 1?

The above is of course known to be true for fields of characteristic 0.

If I is an ideal in $K[G]$, then we set $G(I) = \{x \in G \mid 1 - x \in I\}$.

Lemma. $G(I)$ is a normal subgroup of G. If I_1 and I_2 are two ideals in $K[G]$ with $I_1I_2 = I_2I_1$, then $(G(I_1), G(I_2)) \subseteq G(I_1I_2)$.

Proof. Let $x, y \in G(I)$, $z \in G$. Then

$$1 - x^{-1} = (-x^{-1})(1 - x) \in I$$
$$1 - xy = (1 - x) + (1 - y) - (1 - x)(1 - y) \in I$$
$$1 - x^z = z^{-1}(1 - x)z \in I$$

so $G(I)$ is a normal subgroup. Now let $x \in G(I_1)$ and $y \in G(I_2)$. Then

$$\begin{aligned} 1 - x^{-1}y^{-1}xy &= x^{-1}y^{-1}(yx - xy) \\ &= x^{-1}y^{-1}((1 - y)(1 - x) - (1 - x)(1 - y)) \in I_1I_2. \end{aligned}$$

Let G be a group and let $\gamma^n(G)$ denote the nth term of the lower central series of G. Then $\gamma^1(G) = G$ and $\gamma^n(G) = (\gamma^{n-1}(G), G)$. Now let p be a fixed prime and let $\delta_p^n(G)$ denote the nth term of a central series of G where $\delta_p^1(G) = G$ and $\delta_p^n(G)$ is generated by the commutator $(\delta_p^{n-1}(G), G)$ and by all pth powers of elements of $\delta_p^{(n/p)}(G)$. Here (n/p) is the smallest integer $\geq n/p$.

Let I denote the kernel of the natural map $K[G] \to K[G/G]$ and define the dimension subgroups of G to be $D^n(K[G]) = G(I^n)$.

Lemma. If K has characteristic 0, then $D^n(K[G]) \supseteq \gamma^n(G)$. If K has characteristic p, then $D^n(K[G]) \supseteq \delta_p^n(G)$.

Proof. By induction on n. The case $n = 1$ is clear and in fact $D^1(K[G]) = G$. By the previous lemma $(D^{n-1}(K[G]), G) \subseteq D^n(K[G])$ so certainly $D^n(K[G]) \supseteq \gamma^n(G)$ in characteristic 0 and $D^n(K[G]) \supseteq (\delta_p^{n-1}(G), G)$ in characteristic p. Let K have characteristic p and let $x \in \delta_p^{(n/p)}(G) \subseteq D^{(n/p)}(K[G])$. Then $1 - x \in I^{(n/p)}$ and hence

$$1 - x^p = (1 - x)^p \in I^{(n/p)p} \subseteq I^n.$$

Thus $x^p \in D^n(K[G])$ and the result follows.

Problem 35. Let K have characteristic 0. Is it necessarily true that $D^n(K[G]) = \gamma^n(G)$?

Problem 36. Let K have characteristic p. Is it necessarily true that $D^n(K[G]) = \delta_p^n(G)$?

These are known in certain special cases. For example, Problem (35) is true if G is a free group (see [*31*]) or a torsion free nilpotent group (see [*24*]) and (36) is true if G is a finite p-group (see [*23*]). They are also known for some small values of n.

It is apparent now that many obvious questions about group rings have been answered and many others have been posed. But certainly the nonobvious and subtle questions will prove to be the most interesting. Therefore in the hope of stimulating research in new directions we offer our final

Problem 37. State and prove an interesting result on group rings.

Added in proof. The polynomial identity problem for group rings has now been completely solved. Let p be a prime. We say A is a p-abelian group if A', the commutator subgroup of A, is a finite p-group.

Theorem (Passman [*67*]). Let K be a field of characteristic p and let G be a group.

i. If G has a p-abelian subgroup A of finite index then $K[G]$ satisfies the standard identity of degree $2[G:A] \cdot |A'|$.

ii. If $K[G]$ satisfies a polynomial identity of degree n, then G has a characteristic p-abelian subgroup A of finite index with $[G:A] \cdot |A'|$ bounded by a fixed function of n.

Corollary. Let K have characteristic p. Then $K[G]$ satisfies a polynomial identity if and only if G has a p-abelian subgroup of finite index.

Theorem (Passman [*69*]). Let $K[G]$ satisfy a polynomial identity of degree n. Then $[G:\Delta(G)] \leq n/2$ and $|\Delta(G)'| < \infty$.

Since the proofs here are almost entirely combinatorial these therefore answer Problems 1, 2, 3, 4(i), and 5. In addition the polynomial identity problem for twisted group rings is solved in [*68*].

We remark that $|\Delta(G)'|$ is not bounded by a function of n in the latter theorem. In the former theorem, the bound for $[G:A] \cdot |A'|$ is astronomical. Set

$$a = a(n) = (n!)^2, \qquad b = b(n) = a^{4^{(a+1)!}},$$
$$c = c(n) = (b^4)^{b^4}, \qquad d = d(n) = (n/2)^c.$$

Then

$$[G:A] \cdot |A'| \leq (a + 1)!\,(c + 1)!\,d.$$

Thus only Problems 11 and 12 remain to be answered and of course work should be done to improve the above bound.

There are examples in the literature which show that Problem 8 is not true in all cases. A field K is absolutely algebraic if it is contained in the algebraic closure of a finite field. A group G is polycyclic if it has a finite subinvariant series with cyclic quotients.

Theorem (Hall [*63*], Levic [*65*]). Let K be an absolutely algebraic field and let G be a polycyclic group. Then every irreducible $K[G]$-module is finite dimensional over K.

For example, let

$$G = \langle x, y, z \mid (x, z) = (y, z) = 1, \quad (x, y) = z \rangle.$$

Then G is a torsion free polycyclic group which therefore implies that $K[G]$ is semisimple. Moreover it is easy to see that G has no p-abelian subgroups of finite index so $K[G]$ does not satisfy a polynomial identity. Hence $K[G]$ does not have r.b. n for any n even though by the above theorem all its irreducible representations have finite degree.

REFERENCES

The following list is by no means complete.

1. Amitsur, S. A. The radical of field extensions. *Bull. Res. Council Israel*, **7F** (1957), 1–10.

2. ———. On the semi-simplicity of group algebras. *Michigan Math. J.*, **6** (1959), 251–253.

3. ———. Groups with representations of bounded degree II. *Illinois J. Math.* **5** (1961), 198–205.

4. ———. Generalized polynomial identities and pivital monomials. *Trans. Amer. Math. Soc.* **114** (1965), 210–226.

5. ———. A noncommutative Hilbert basis theorem and subrings of matrices. *Trans. Amer. Math. Soc.* **149** (1970), 133–142.

6. Amitsur, S. A., and Levitzki, J. Minimal identities for algebras. *Proc. Amer. Math. Soc.* **1** (1950), 449–463.

7. Artin, E. Galois Theory. Notre Dame Mathematics Lectures, No. 2, Notre Dame, Indiana, 1959.

8. Bass, H. Finitistic dimension and a homological generalization of semi-primary rings. *Trans. Amer. Math. Soc.* **95** (1960), 466–488.

9. Bovdi, A. A. Group rings of torsion free groups (in Russian). *Sibirsk. Mat. Z.* **1** (1960), 555–558.

10. Brauer, R. Zur Darstellungstheorie der Gruppen endlicher Ordnung I. *Math. Z.*, **63** (1956), 406–444.

11. Connell, I. G. On the group ring. *Canad. J. Math.*, **15** (1963), 650–685.

12. Green, J. A., and Stonehewer, S. E. The radicals of some group algebras. *J. Algebra*, **13** (1969), 137–142.

13. Golod, E. S. On nil algebras and finitely approximable groups (in Russian). *Izv. Akad. Nauk SSSR Ser. Mat.*, **28** (1964), 273–276.

14. Golod, E. S., and Shafarevitch, I. R. On towers of class fields (in Russian). *Izv. Akad. Nauk SSSR Ser. Mat.*, **28** (1964), 261–272.

15. Hall, P. Finiteness conditions for soluble groups. *Proc. London Math. Soc.*, **4** (1954), 419–436.

16. Herstein, I. N. Noncommutative Rings. Carus Mathematics Monographs, No. 15, Math. Assoc. Amer., 1968.
17. Higman, D. G. Modules with a group of operators. *Duke Math. J.*, **21** (1954), 369–376.
18. Higman, G. The units of group rings. *Proc. London Math. Soc.*, **46,** (2) (1940), 231–248.
19. Isaacs, I. M., and Passman, D. S. Groups with representations of bounded degree. *Canad. J. Math.*, **16** (1964), 299–309.
20. ———. A characterization of groups in terms of the degrees of their characters. *Pacific J. Math.*, **15** (1965), 877–903.
21. ———. A characterization of groups in terms of the degrees of their characters II. *Pacific J. Math.*, **24** (1968), 467–510.
22. Jacobson, N. Structure of Rings. Amer. Math. Soc. Colloquium Publications, Vol. 28, Providence, Rhode Island, 1956.
23. Jennings, S. A. The structure of the group ring of a p-group over a modular field. *Trans. Amer. Math. Soc.*, **50** (1941), 175–185.
24. ———. The group ring of a class of infinite nilpotent groups. *Canad. J. Math.*, **7** (1955), 169–187.
25. Kaplansky, I. Rings with a polynomial identity. *Bull. Amer. Math. Soc.*, **54** (1948), 575–580.
26. ———. Groups with representations of bounded degree. *Canad. J. Math.*, **1** (1949), 105–112.
27. ———. Problems in the theory of rings. NAS-NRC Publ. 502, Washington, 1957, 1–3.
28. ———. Fields and Rings. Chicago Lectures in Mathematics, Univ. Chicago Press, Chicago, 1969.
29. ———. "Problems in the theory of rings" revisited. *Amer. Math. Monthly*, **77** (1970), 445–454.
30. Lambek, J. Lectures on Rings and Modules. Blaisdell, Waltham, Massachusetts, 1966.
31. Magnus, W. Bezichungen zwischen Gruppen und Idealen in einem speziellen Ring. *Math. Ann.*, **111** (1935), 259–280.
32. Martindale, W. S., III. Prime rings satisfying a generalized polynomial identity. *J. Algebra*, **12** (1969), 576–584.
33. Montgomerty, M. S. Left and right inverses in group algebras. *Bull. Amer. Math. Soc.*, **75** (1969), 539–540.
34. Osima, M. Note on blocks of group characters. *Math. J. Okayama Univ.*, **4** (1955), 175–188.
35. Passman, D. S. Nil ideals in group rings. *Michigan Math. J.*, **9** (1962), 375–384.
36. ———. Character kernels in discrete groups. *Proc. Amer. Math. Soc.*, **17** (1966), 487–492.
37. ———. On the semisimplicity of modular group algebras. *Proc. Amer. Math. Soc.*, **20** (1969), 515–519.
38. ———. On the semisimplicity of modular group algebras II. *Canad. J. Math.*, **21** (1969), 1137–1145.
39. ———. Central idempotents in group rings. *Proc. Amer. Math. Soc.*, **22** (1969), 555–556.

40. ———. Radicals of twisted group rings. *Proc. London Math. Soc.*, **20** (1970), 409–437.
41. ———. On the semisimplicity of twisted group algebras. *Proc. Amer. Math. Soc.*, **25** (1970), 161–166.
42. ———. Twisted group algebras of discrete groups. *Proc. London Math. Soc.*, **21** (1970) 219–231.
43. ———. Radicals of twisted group rings II. *Proc. London Math. Soc.*, **22** (1971) 633–651.
44. ———. Linear identities in group rings. *Pacific J. Math.*
45. ———. Linear identities in group rings II. *Pacific J. Math.*
46. ———. Idempotents in group rings. *Proc. Amer. Math. Soc.*
47. Posner, E. C. Prime rings satisfying a polynomial identity, *Proc. Amer. Math. Soc.*, **11** (1960), 180–184.
48. Rickart, C. The uniqueness of norm problem in Banach algebras. *Ann. Math.*, **51** (1950), 615–628.
49. Rudin, W., and Schneider, H. Idempotents in group rings. *Duke Math. J.*, **31** (1964), 585–602.
50. Smith, M. On group algebras. *Bull. Amer. Math. Soc.*, **76** (1970), 780–782.
51. ———. Thesis, Univ. of Chicago, 1970.
52. Stonehewer, S. E. Group algebras of some torsion-free groups. *J. Algebra*, **10** (1969), 143–147.
53. Villamayor, O. E. On the semisimplicity of group algebras. *Proc. Amer. Math. Soc.*, **9** (1958), 621–627.
54. ———. On the semisimplicity of group algebras II. *Proc. Amer. Math. Soc.*, **10** (1959), 27–31.
55. ———. On weak dimension of algebras. *Pacific J. Math.*, **9** (1959), 941–951.
56. Von Neumann, J. On regular rings. *Proc. Nat. Acad. Sci. U.S.A.*, **22** (1936), 707–713.
57. Wallace, D. A. R. The Jacobson radicals of the group algebras of a group and of certain normal subgroups. *Math. Z.*, **100** (1967), 282–294.
58. ———. Some applications of subnormality in groups in the study of group algebras. *Math. Z.*, **108** (1968), 53–62.
59. ———. On commutative and central conditions on the Jacobson radical of the group algebra of a group. *Proc. London Math. Soc.*, **19** (1969), 385–402.
60. ———. The radical of the group algebra of a subgroup, of a polycyclic group and of a restricted *SN*-group. *Proc. Edinburgh Math. Soc.*
61. Woods, S. M. On perfect group rings. *Proc. Amer. Math. Soc.*
62. Zalessky, A. E. On group rings of solvable groups (in Russian). *Izv. Akad. Nauk. BSSR*, **2** (1970), 13–21.
63. Hall, P. On the finiteness of certain soluble groups, *Proc. London Math. Soc.*, (3) **9** (1959), 595–622.
64. Herstein, I. N. Notes from a ring-theory conference.
65. Levic, E. M. On a problem of P. Hall (in Russian), *Dokl. Akad. Nauk. SSSR*, **188** (1969), 1241–1243.
66. Neumann, B. H. Groups covered by permutable subsets, *J. London Math. Soc.*, **29** (1954), 236–248.
67. Passman, D. S. Group rings satisfying a polynomial identity, *J. Algebra.*
68. ———. Group rings satisfying a polynomial identity II, *Pacific J. Math.*

69. ———. Group rings satisfying a polynomial identity III, *Proc. Amer. Math. Soc.*

70. ———. Minimal ideals in group rings, *Proc. Amer. Math. Soc.*

71. Rosen, M. I. The Jacobson radical of a group algebra, *Mich. Math. J.*, **13** (1966), 477–480.

72. Rosenberg, A. E. On the semisimplicity of the group algebra, Ph.D thesis, Brown Univ., 1969.

73. Wiegold, J. Groups with boundedly finite classes of conjugate elements, *Proc. Roy. Soc. London*, *Ser. A.*, **238** (1957), 389–401.

NOTATION

Set theoretic:

$\|S\|$	the size of the set S
$\supseteq$	weak inclusion
$>$	strong inclusion
$\varnothing$	the empty set

Number theoretic:

$a \mid b$	a divides b
$[a]$	the greatest integer in a
$\mathfrak{A}(n)$	$(n!)^{n\cdot(n!)^2}$
$\mathfrak{A}(n, m)$	$(n!)^{2m\cdot(n!)^2}$
$(F:K)$	the degree of the field extension F/K
t.d. (F/K)	the transcendence degree of the field extension F/K

In the group G:

$\mathbf{Z}(G)$	the center of G
$\mathbf{C}_G(x)$	the centralizer of $x \in G$ in G
$\mathbf{N}_G(H)$	the normalizer of the subgroup $H \subseteq G$
$[G:H]$	the index of H in G
Hx	a coset of H in G
G/H	the quotient group of G by its normal subgroup H

$\langle \cdots \rangle$	the subgroup of G generated by the elements and subsets indicated
$x^y = y^{-1}xy$	the conjugate of x by y
$(x, y) = x^{-1}y^{-1}xy$	the commutator of $x, y \in G$
(H_1, H_2)	$\langle (h_1, h_2) \mid h_i \in H_i \rangle$, the commutator of two subgroups of G
$G' = (G, G)$	the commutator subgroup of G
$\Delta(G)$	$\{x \in G \mid [G : C_G(x)] < \infty\}$, the F.C. subgroup of G
$\Delta^+(G)$	the torsion subgroup of $\Delta(G)$
$\Delta^p(G)$	the subgroup of $\Delta^+(G)$ generated by all elements whose order is a power of p
$\mathbf{S}(G)$	the subgroup of G generated by all solvable normal subgroups of G

Group constructions:

$G \wr H$	the Wreath product of G by H
$G \times H$	the direct product of G and H
$G \times_\sigma H$	the semidirect product of G by H
ΠG_ν	the weak direct product of the groups G_ν (each element has only finitely many nonidentity projections)

In the ring R:

I	an ideal (two-sided unless otherwise stated)
$[\alpha, \beta] = \alpha\beta - \beta\alpha$	The Lie product of $\alpha, \beta \in R$
$[R, R]$	the Lie product of R with R, that is the additive subgroup spanned by all $[\alpha, \beta]$
JR	the Jacobson radical of R
NR	the nilpotent radical of R
$Q(R)$	a complete ring of quotients of R, if it exists
R_m	the ring of $m \times m$ matrices over R
K	a field
$K[\zeta_1, \zeta_2, \ldots, \zeta_n]$	the polynomial ring over K in the noncommuting variables $\zeta_1, \zeta_2, \ldots, \zeta_n$
$[\zeta_1, \zeta_2, \ldots, \zeta_n]$	$s_n(\zeta_1, \zeta_2, \ldots, \zeta_n)$, the standard polynomial of degree n
V, W	right R-modules
End V	the ring of endomorphisms of V
$\mathrm{Hom}_R(V, W)$	the set of R-homomorphisms from V to W
$R \otimes_K S$	the tensor product of the two K-algebras R and S
$V \otimes_K W$	the tensor product of the two K-modules V and W
ΣV_ν	the weak direct sum of the modules V_ν (each element has only finitely many nonzero projections)

In the group algebra $K[G]$

$\alpha = \Sigma_{x \in G} a_x \cdot x$	a typical element with $a_x \in K$
$\mathrm{tr}\, \alpha = a_1$	the trace of α, that is its coefficient of $1 \in G$
Supp α	$\{x \in G \mid a_x \neq 0\}$, the support of α
$\theta : K[G] \to K[\Delta(G)]$	the natural projection
$\theta^+ : K[G] \to K[\Delta^+]$	the natural projection
$\varphi : K[G] \to K[\mathbf{Z}(G)]$	the natural projection
χ_V	the group character for the finite dimensional $K[G]$-module V

INDEX